Acclaim for

SHADOWS
in the
VINEYARD

"A rare book that transcends the narrow interests of wine lovers."
—*The New York Times*, named a Best Wine Book of 2014

"In the end, the book turns out to be less about a dastardly crime than about the powerful mysticism surrounding Burgundy's wine culture... Powerful."
—*The New York Times Book Review*

"SHADOWS IN THE VINEYARD details the tale of a threat so bizarre it seems unreal: one deluded man's attempt at a big score by threatening a true culinary treasure... the prose is evocative... and captures the gravitas of a crime with deep cultural resonance without becoming wonky." —*San Francisco Gate*

"Books I devoured in the last few weeks impressed me mightily:... SHADOWS IN THE VINEYARD, a nonfiction blending of several genres—culinary, travel, history, true crime by Maximillian Potter, who's an enviably nimble writer."
—Frank Bruni, *New York Times*

"Mr. Potter... tells the story quite vividly—and the book delivers a happy ending as well." —*Wall Street Journal*, Great Reads

"A gripping crime drama more creative than most procedurals, and Potter does excellent working in fleshing out both the involved players and the historical context of the Burgundy region and its oenophiles."

—*The Daily Beast*

"[Potter] places the crime in a broad, rich, historical and cultural context that is engaging"

—*USA Today*

"A whodunit with a culprit worthy of a Woody Allen film, Potter's first book is a rich study of a cinematic crime and bona fide page-turner…Even the most devout teetotaler will have a hard time putting this one down."

—*Publishers Weekly* (starred review)

"A satisfying crime story expanded into a love letter to a great French vineyard."

—*The Dispatch*

"Remarkable…Through his immersive reporting, Potter gives us a privileged look at today's Burgundy."

—Terroirist

"Inspired by a daring crime that a lesser writer might reduce to police procedural, Maximillian Potter has opened a portal into a fabled world unknown to outsiders. The story he so compellingly recounts in SHADOWS IN THE VINEYARD breathes, like the hallowed wine at its heart, with life and history and wonder."

—Benjamin Wallace, author of
The Billionaire's Vinegar

"Aubert de Villaine is the legend behind the legend. DRC is the most celebrated wine on the planet and the place where the alchemy of the soul of the earth, combined with the elements, as well as the knowledge and wisdom of craftsmen, are united to create this magical nectar. But out of nowhere, the dark side interfered and this inconceivable thriller began. This riveting story, where good ultimately triumphs, instills a renewed appreciation of the Côte d'Or region, its people, and the passion that is the fortitude behind this incredible wine."

—Eric Ripert, chef and co-owner,
Le Bernardin, author of *Avec Eric*

"A gripping, real-life mystery and an intimate portrait of one of the world's great wine-makers as he battles the man bent on destroying five centuries of greatness. Maximillian Potter has always been an outstanding reporter and now he reveals the fascinating story of France's legendary vineyard, Domaine Romain Conti."

—Michael Hainey, author of
After Visiting Friends

"An arch-criminal clicks on his headlamp in his underground lair and instantly, you're hooked. SHADOWS IN THE VINEYARD is non-fiction at its nail-biting best, a literary true-crime thriller that plunges you into the manhunt to apprehend—and understand—a mysterious villain who set out to destroy the most treasured wines in the world. SHADOWS IN THE VINEYARD is so full of bizarre twists and one-of-a-kind characters that if you think you know what's coming next, just wait till you turn the page."

—Christopher McDougall,
author of *Born to Run*

"Maximillian Potter has taken a sinister plot and woven an intriguing story around the most revered wine estate in the world with the most respected winemaker at its helm. Through this event he has painted a colorful tableau filled with fascinating historical evidence on why the Domaine de la Romanée-Conti, the terroir of Burgundy, and the culture of Burgundy, are among the most treasured and special sites on the planet. Bravo."

—Daniel Boulud, James Beard
Award–winning chef, and Daniel Johnnes,
James Beard Award–winning sommelier

SHADOWS *in the* VINEYARD

THE TRUE STORY OF THE PLOT TO POISON THE WORLD'S GREATEST WINE

Maximillian Potter

TWELVE

NEW YORK BOSTON

Twelve
Grand Central Publishing
Hachette Book Group
1290 Avenue of the Americas
New York, NY 10104

www.HachetteBookGroup.com

Printed in the United States of America

RRD-C

Originally published in hardcover by Hachette Book Group.

First trade edition: July 2015
10 9 8 7 6 5 4

Twelve is an imprint of Grand Central Publishing.
The Twelve name and logo are trademarks of Hachette Book Group, Inc.

Library of Congress Cataloging-in-Publication Data

Potter, Maximillian.
Shadows in the vineyard : the true story of a plot to poison the world's greatest wine / Maximillian Potter. — First edition.
pages cm
ISBN 978-1-4555-1610-0 (hardback) — ISBN 978-1-4555-1608-7 (ebook) — ISBN 978-1-61113-203-8 (audio download) 1. Wineries— France—Burgundy—Case studies. 2. Viticulture—France—Burgundy—Case studies. 3. Grapes—France—Burgundy—Herbicide injuries—Case studies. 4. Crime—France—Burgundy—Case studies. I. Title.
TP553.P63 2014
364.16'4—dc23

2014015787

ISBN 978-1-4555-1609-4 (pbk.)

For Lori and our enfants, True and Jack

WHAT KIND OF MAN THE CELLARER OF THE MONASTERY SHOULD BE

1) As cellarer of the monastery should be chosen from the community, one who is sound in judgment, mature in character, sober, not a great eater, not self-important, not turbulent, not harshly spoken, not an off-putter, not wasteful

2) but a God-fearing man, who will be a father to the whole community

3) He is to have charge of all affairs

10) He must regard the chattels of the monastery and its whole property as if they were sacred vessels of the altar

[Chapter 31 of the Benedictine Rule, as posted in English inside the Burgundy's Abbey Notre Dame de Cîteaux]

Contents

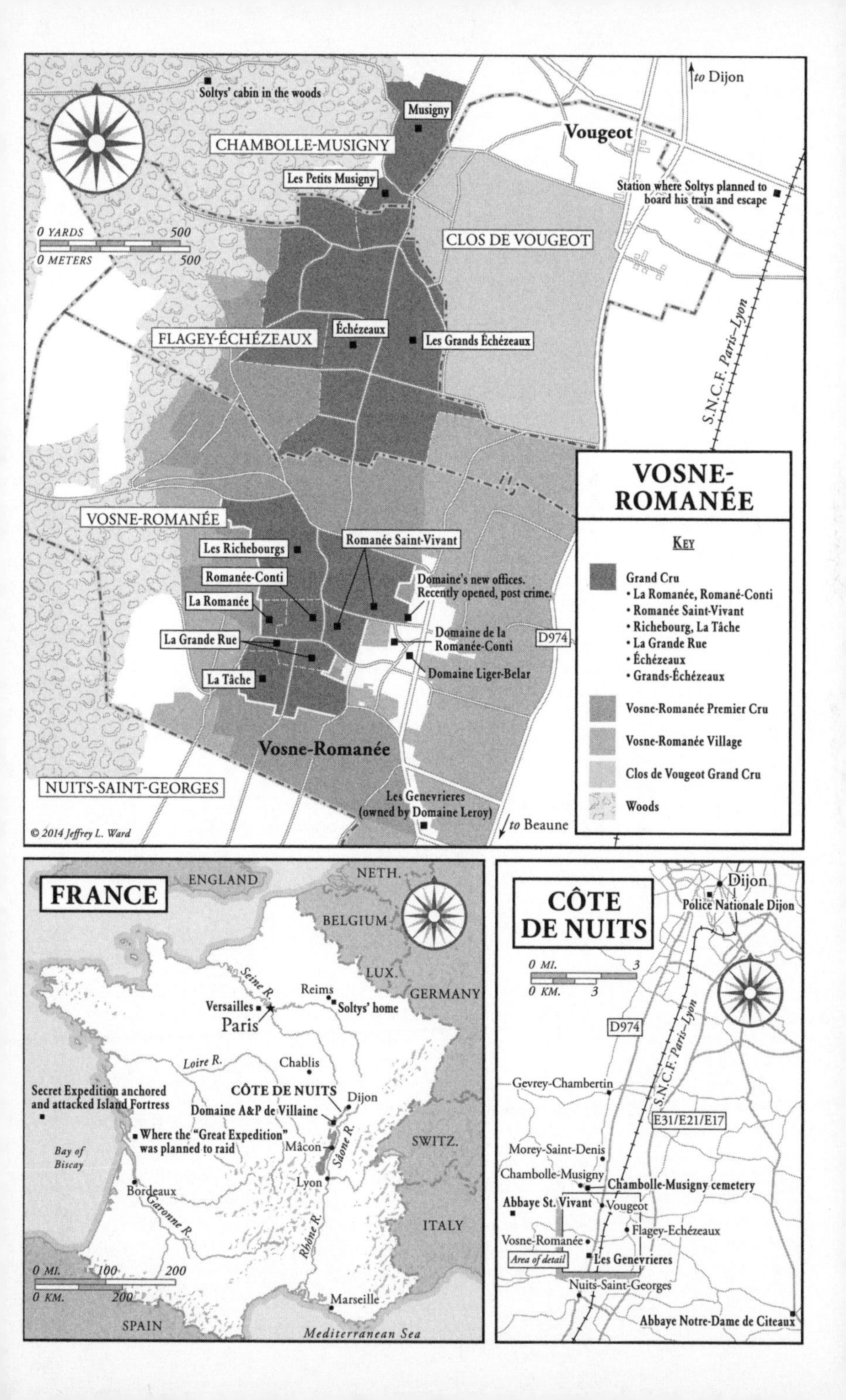
Soltys' cabin in the woods
to Dijon
Musigny
CHAMBOLLE-MUSIGNY
Vougeot
Les Petits Musigny
Station where Soltys planned to board his train and escape
0 YARDS 500
0 METERS 500
CLOS DE VOUGEOT
Échézeaux
FLAGEY-ÉCHÉZEAUX
Les Grands Échézeaux
S.N.C.F. Paris–Lyon
VOSNE-ROMANÉE
VOSNE-ROMANÉE
Key
Les Richebourgs
Romanée Saint-Vivant
Romanée-Conti
Domaine's new offices. Recently opened, post crime.
La Romanée
Domaine de la Romanée-Conti
D974
La Grande Rue
Domaine Liger-Belar
La Tâche
Vosne-Romanée
NUITS-SAINT-GEORGES
Les Genevrieres (owned by Domaine Leroy)
to Beaune
© 2014 Jeffrey L. Ward
Grand Cru
• La Romanée, Romané-Conti
• Romanée Saint-Vivant
• Richebourg, La Tâche
• La Grande Rue
• Échézeaux
• Grands-Échézeaux
Vosne-Romanée Premier Cru
Vosne-Romanée Village
Clos de Vougeot Grand Cru
Woods
FRANCE
ENGLAND
NETH.
BELGIUM
LUX.
GERMANY
Seine R.
Reims
Versailles
Soltys' home
Paris
Loire R.
Chablis
CÔTE DE NUITS
Dijon
Secret Expedition anchored and attacked Island Fortress
Domaine A&P de Villaine
Where the "Great Expedition" was planned to raid
Mâcon
Saône R.
SWITZ.
Bay of Biscay
Lyon
Bordeaux
Garonne R.
Rhône R.
ITALY
0 MI. 100 200
0 KM. 200
Marseille
SPAIN
Mediterranean Sea
CÔTE DE NUITS
Dijon
Police Nationale Dijon
0 MI. 3
0 KM. 3
D974
S.N.C.F. Paris–Lyon
Gevrey-Chambertin
E31/E21/E17
Morey-Saint-Denis
Chambolle-Musigny
Chambolle-Musigny cemetery
Abbaye St. Vivant
Vougeot
Flagey-Echézeaux
Vosne-Romanée
Area of detail
Les Genevrieres
Nuits-Saint-Georges
Abbaye Notre-Dame de Citeaux

CHAPTER 1

The Grand Monsieur

The sun over Burgundy's seemingly endless expanse of richly green vineyards belonged to late summer. What few clouds there were, were fantastical, fat, and luminous—giant dollops of silver and white acrylic paint that had not yet finished drying onto God's vast canvas of sky. Plush canopies of leaves on the tens of thousands of vines fluttered in breezes so faint that if not for the subtle sway it would have appeared there wasn't any breeze at all. Chirping sparrows swooped every which way, as if they'd spent the night drinking from an open barrel in one of the nearby *cuveries*. With the gentle rise and fall of the terrain, the vineyards resembled a slow rolling ocean of unpredictable currents.

The temperature that September morning in 2010, in the Côte d'Or region, which is the heart of Burgundy, and for many serious wine collectors the only part of Burgundy that matters, was already well on its way to sweltering. The humidity was as present as a coastal mist. Soon, the workers would spill from the villages to tend to the vines. The *enjambeurs* would arrive: The spider-shaped tractors, with their high tires to easily traverse the meticulously ordered vine rows, would scurry about dodging the tourists bicycling along the narrow ribbons of dirt road between the vineyards. For the moment, however, the landscape was quiet; as far as the eye could see the only person among the vines was the *Grand Monsieur*.

Dressed in shades of khaki—even his wide-brimmed, cloth hat—seventy-one-year-old Aubert de Villaine walked in the parcel called Romanée-St.-Vivant. Tall and thin, he waded through the vines as he had done for more than four decades: in bursts of long strides, arms out slightly from his sides, palms skimming the vine tops.

Every so often he would stop, fish the handkerchief from his pocket, wipe the perspiration from his brow, and look about. Monsieur de Villaine knew that everything and nothing was unfolding before his eyes, and that it was his challenge to determine which was the everything and which was the nothing—to find the clues in nature's mystery.

At moments like this, surrounded by the sublime splendor of the vineyards before the harvest, the Grand Monsieur sometimes thought of the French masters—Pissarro, Renoir, Monet. He suspected they would have appreciated Burgundy and understood his work.

"One must have only one master—nature," Pissarro had said. Renoir had put it this way: "You come to nature with your theories and she knocks them all flat." And Monet—ah, Monet. Was it any wonder he described it best of all? "A landscape hardly exists at all as a landscape because its appearance is changing in every moment. But it lives through its ambiance, through the air and light, which vary constantly."

Though Monsieur de Villaine would have insisted he was unworthy of such a comparison, he had much in common with

Monet. When Monet first picked up his brush he saw and painted the natural world in pieces. He put the water here and the sky there; the field went here; flowers and trees went here, here, and there. Each was an element unto itself, existing almost independent of its surroundings, as if, just like that, any one of the elements could have just as easily been placed in another scene, transported to another painting.

As he matured, however, Monet's work became less technical and more organic, spiritual. He came to understand nature's power. It was as if one day, while standing alone on the banks of that pond covered in water lilies, Monsieur Monet discovered a crease in the universe, pulled it open like curtains, stepped inside, and turned and viewed the world from another dimension—from a perspective that allowed him to see the interconnectedness of it all, to see the light and air, and the flicker and flow of energy among all natural things.

It was then that Monet began to make the invisible visible. The lines he had once thought defined and separated some natural order dissolved into a liquefied oneness, filling the canvas for others to drink in, and, if only for a few moments, to experience the divine.

This was what the Grand Monsieur labored to do. Only with grapes. His life had been dedicated to transcending the technical and vinifying nature's invisible energy.

As he studied the masterpiece of the landscape around him, the Grand Monsieur prayed for a sign. He prayed although he wasn't as confident in the power of prayer as he once had been. Because of recent horrific events and the possibility of unsettling outcomes, the Grand Monsieur had begun to question God's very existence.

You wouldn't have been able to detect his fermenting

anxieties just by looking at him. Or maybe you could have. If you were among the very few he trusted to know him well enough, and you happened to glimpse him in a moment like this—his long, weathered face and forlorn brown eyes—when he thought no one was looking at him, when he thought he was alone and could be himself.

Then again, it had been so long since Monsieur de Villaine had known what that was like: a private moment, unto himself. There was no himself. Only the tangle of what he represented: the vines, the families, the Domaine, Burgundy, France—the storied legacy of countless holy men and one unholy prince. A legacy subjected to the currencies of markets too often ignorant of how to truly appreciate a bottle of Burgundy, and that were instead driven by the whims of buyers who were obsessed with the bangs of auction gavels and status-symbol trophy bottles.

For the longest time Monsieur de Villaine had wanted no parts of any of it. He had resisted. It would be fair to say he had fled Burgundy's vines. But crawling into a blackness that he believed was certain death, riding horseback into the starry night of the American West, repeatedly enduring the heartache of the unborn—well, these things have a way of altering a man.

Over time, like the best Pinot Noirs, within the bottle of his skin Monsieur de Villaine's composition had become something other than what it had been. He matured. He came to accept and to appreciate what had always been his destiny—caring for his *enfant* vines, and producing the most magnificent and most misunderstood wine in the world.

His *employés* referred to him as the Grand Monsieur. The moniker signified their respect. A recognition of his grace and kindness. Monsieur de Villaine put up a reserved front, but his people knew it was a façade, his way of protecting the Domaine and also his own heart, broken several times over and patched together, it seemed, with rose petals.

Years ago, when one of their beloved fellow workers, distraught over a lost love, was found hanging from a rafter in the winery, it was the Grand Monsieur to whom they turned for guidance. The workers gathered in the winery and bowed their heads as he led them in prayer and reminded them that the Lord indeed works in mysterious ways. He told them this was part of God's plan. He petitioned them to have faith, to believe.

When the torrents of rain came and lasted for days and drowned their scheduled vineyard tasks, as was often the case in mid- to late summer just before the *vendange*, it was the Grand Monsieur who assuaged their concerns. Although during such times he more than anyone else worried that they would fall behind or the crop might be lost, he exuded a serenity; he reassured his crew that when the skies would clear they would be able to complete the work. In due time, he would tell his men when their partners, God and Mother Nature, were ready. Have faith, he would say to them. Believe.

Although the Grand Monsieur presided over the Domaine that was above all other Burgundian domaines, a national treasure—a "cathedral" of a winery, as a French official in Paris had once put it—and a winery that had made him one of the wealthiest men in France, Monsieur de Villaine carried himself with equal parts dignity and humility. He emanated gratitude and took nothing for granted. He was ever mindful of the time when the Domaine's wines made no profit at all, and he never lost sight of the fact that such a period could easily come again.

He was often one of the first to arrive at the Domaine in the morning, in his silver Renault station wagon, and he was among the last to leave. His back was just as sore as theirs; his hands just as calloused. He was the kind of Grand Monsieur who once, when his wife, Pamela, asked him to travel into Paris to meet her at a party hosted by an American starlet, he lingered at the "farm," as he called the world's greatest Domaine, for as long as he could, and then only grudgingly attended the soirée. He arrived late, as one of his family members recalled the evening, dressed in his khaki farmhand clothes.

During the long days of the *vendange*, the Grand Monsieur made sure his pickers were paid much better than the other domaines' crews were paid; he contracted a locally renowned chef to prepare their meals. Rather than ensconce himself in an air-conditioned office, he opted to be in the *cuverie*, or in the vineyards for the clipping and sorting, often inquiring about his employees' welfare and their families. He asked, his people knew, because he cared. As far as the Grand Monsieur was concerned, anyone who worked at the Domaine was family; they grasped they were part of something very special.

One day during a harvest, as the Grand Monsieur's *vendangeurs* picked the Domaine's vines in the parcel called Richebourg, one of his workers approached me. Squat, husky, with a nose that looked as if it got smashed crooked and flat by a barroom one-two wallop. He wore shorts, a white T-shirt tank top, work boots, and a skullcap. A cigarette dangled from the side of his mouth. He struck me as someone who would be more at home as a stevedore on the docks heaving bags of coffee beans or bundles of bananas rather than given over to the painstaking, delicate detail work of harvesting tiny bundles of berries.

He shouldered a backpack pannier and was tasked with transporting the picked fruit from the vineyard to a nearby flatbed trailer. "Pannier! Pannier!" The pickers called for him when their harvesting baskets were filled. For days as he worked I'd watched him stealing peeks at some of the female pickers as they bent over using their wire-cutter-like secateurs to clip off the clusters of the Pinot Noir. At least once he'd caught me watching him. I'd gotten the impression this Monsieur Pannier didn't care to be observed. I thought he was going to tell me as much when he approached me.

"Can I tell you something?" He spoke to me in English. By then I'd been around the Domaine for two harvests and for enough months that everyone at the Domaine knew I spoke very little French. He stood with his face inches from my face. I could smell his sweat and the nicotine on his breath.

"*Bien sûr*," I answered. Of course. I tried to use what little I knew of the language.

He flicked his cigarette onto a nearby ribbon of road. I reassured myself that it was unlikely he would pick a fight with me here, in front of everyone.

"The big boss," he said, nodding in the direction of Monsieur de Villaine. The big boss was well out of earshot, over on the back of the flatbed, alone, sorting through the grapes that had been picked and carefully poured from the panniers into the plastic crates that would be transported to the *cuverie*.

"*Oui?*" I said.

"Son cœur est dans la terre."

Pannier could see I was trying to process his French. He knelt down in front of me. Genuflected was more like it. He took his right hand and pressed his palm flat to his chest, over his heart. He looked up at me, his eyes locking on to mine, to make sure I was watching his gesture.

"*Son cœur*," he said.

"His heart?"

"*Oui, son cœur*." He removed the hand from his chest and pressed it into the soil. "*Est dans la terre*."

"Is in the earth?"

"*Oui*." He stood. He looked into me until he was satisfied I understood.

He softly punched my shoulder and said it again:

"*Monsieur de Villaine, le Grand Monsieur, son cœur est dans la terre*."

With that, Monsieur Pannier smiled in the big boss's direction and walked off back into the vines to see the sights and wait for his next load of fruit.

On the quiet morning in mid-September 2010, as the Grand Monsieur walked through the vines of Romanée-St.-Vivant and looked to the sky, he searched for clues that would help him determine when to begin another year's *vendange*.

Off and on, for centuries, ruling aristocrats and government officials had set the date for the start of the harvest for all of Burgundy. Typically, and most unfairly, this *ban de vendange* corresponded with the wishes of the wealthiest owners of the finest vineyards, which produced the highest quality grapes. In theory, the policy of a unified harvest period made sense, as the harvest took over the entire region. Horse-drawn wagons filled with grapes on their way to the wineries clogged the rural roads and tight city streets. Businesses closed, willingly or otherwise, to allow friends and family of vignerons to pick and sort. But in the way that mattered most, for bureaucrats to choose when the harvest would begin for all was inherently flawed policy.

Only the vigneron who tends his vines knows when his berries are ready. Only the vine farmer himself knows that the grapes growing in one section of his vineyard, say, where there tends to be more exposure to sunlight and wind, will mature faster than the berries in another section of that same parcel. An east-facing slope of vines likely gets more of the hot midday sun. And so on.

Then there's the myriad farming techniques. Each Burgundian grower has his own way of doing things—a hybrid of science and metaphysical voodoo, informed by tradition and faith, and, of course, viticulture. So many nuances. In the end, no one understands the contours of a parcel of vines better than its vigneron. The way Mark Twain's riverboat captains knew the secret shoals of the Mississippi. The way a husband understands the curves and mysteries of his beloved's form.

France is relatively small country, eight thousand square miles smaller than the geometrically similar state of Texas. The Burgundy region, the mostly pastoral countryside to the southeast of Paris and comprising four "departments"—the Yonne, Nièvre, Saône-et-Loire, and the Côte d'Or—represents only about one-twentieth of the country. The roughly forty-mile-long by three-mile-wide corridor of Côte d'Or wine country, which stretches from the city of Dijon to just south of the city of Beaune, is little more than a wrinkle in the universe.

Yet within that wrinkle, the temperature and the terrain vary dramatically. The Côte d'Or is divided into two regions: Vineyards in the south belong to the Côte de Beaune, and the vineyards in the north are in the Côte de Nuits, which at the time was where all but one parcel of Monsieur de Villaine's vines grew. Although the two *côtes* (literally, slopes) are in such intimate proximity, they may as well be on different planets when it comes to late summer weather. So the officials in Beaune ultimately surrendered to the reality that the decision of when to harvest should rest where it does now, with the vignerons themselves.

When Monsieur de Villaine walked his vines he would sometimes picture the prehistoric ocean that covered this part of France. Visions of ancient fish floated like sunspots before his eyes. He watched the creatures swim, then, as the earth's crust moved apart and came together, pushing up mountains and cracking off faces of cliff—as the ocean receded—he watched as the sea creatures fossilized, atomized, sprinkled down, and vanished into the soil. He saw the holy ghosts enter the wild land—the monks in their pointed hoods cutting away brush, raking the earth, then kneeling and putting the earth in their mouths, and then marrying their vines to the soil.

Monsieur de Villaine sensed the energy in the veins of the earth around him, an energy that would infuse the Burgundy wines that King Charlemagne had so very long ago declared worthy to be consecrated the blood of Christ. The Grand Monsieur imagined the princely namesake of his Domaine pacing these vines, ensuring his parcels were not too densely planted, insisting that quality never be compromised in favor of quantity. Of course, too, the Grand Monsieur would see himself as a boy in these vines, disinterested and trailing behind his own father and grandfather.

Like virtually all Burgundians, the de Villaines were Catholic. The Grand Monsieur had spent a fair amount of his life in churches. He likened the many thousands of vineyards of Burgundy to the shards of a stained glass window. Thousands upon thousands of parcels divided, seemingly without rhyme or reason, and within those parcels, a range of asymmetrical *climats* that were at once unto themselves and yet exquisitely pieced together into a meticulously engineered, breathtaking whole.

So far, the growing season of 2010 had brought much rain and humidity. Which could mean disaster for the Pinot Noir. "Pinots" are so named because the clusters of this grape varietal resemble a pinecone. Just as the structure of a pinecone is as dense as it is delicate, the Pinot grapes grow in tight bunches that leave little room for the flow of air between the berries. Under the shade of the canopies of leaves, within the tight cone-shaped clusters of 2010, because of the humidity, moisture had set in.

If Monsieur de Villaine timed his harvest too late, rot and mildew might eat away the grape skins. The botrytis fungus might render the grapes so many white, dusty cadavers that would turn to dust during picking. It was bittersweet irony that as the berries matured to peak ripeness and sugar levels, just when they would have the best to offer, they were simultaneously decaying, soon fit only to be left on the vine, to fall off and die into the soil.

At this stage of his life this viticultural reality was something Monsieur de Villaine understood quite well. Many who knew him, or thought they knew him, whispered that he was beginning to appear frail and often seemed fatigued; that his vision and instincts were not quite as sharp as they had been. There was the story circulating in the vineyards that driving home one evening he had struck a young girl on a bicycle. It was nothing serious. And of course, according to the talk, the Grand Monsieur felt terrible and had visited the young girl in the hospital. In short, people had begun to wonder how many vintages Monsieur de Villaine had left in him.

He was aware of the talk. He pretended not to hear it or care, but he did care. Such whispers raised the question of whether he was leaving himself too long on the vine.

Truth be told, there were times when he thought of the talk and it caused him to doubt himself. More often than not, when he considered the gossip it emboldened him. He would shrug and blow the air from his cheeks—as the French like to do—and he would tell himself that he still had much of his best to give.

Besides, he thought, *what was the alternative for the Domaine?* Whenever the question of his successor crept into his head he told himself he had more pressing matters to resolve, like now, the decision of when to harvest. He persuaded himself to believe that on the matter of his retirement and his heir apparent, the longer he waited the better.

—

The soft September breezes that rustled the waist-high leaves surrounding him were welcome. The winds dried the moisture, combated the fungi, and prolonged those last critical days of ripening, enabling the grapes to reach maximum sugar level and balance; giving them just that much more time to swell on the vine like so many sweet supernovas, which in turn, so went the hope, would infuse the wine with marvelous flavor.

In that September morning's air, though, Monsieur de Villaine sensed the sort of stealthy humid heat that he had come to learn often foreshadowed violent rainstorms, perhaps even hail. This raised more questions that required immediate consideration: Were storms imminent? At what pace were the humidity and rain spreading rot on his fruit? Could he give his berries more time on the vines or did he need have his vineyard manager, Nicolas Jacob, call in the pickers?

The Grand Monsieur wiped his brow with his handkerchief.

The sound of the rustling leaves reminded him of the soft winds that blow across the tiny whitecaps on his favorite fly-fishing spot, the Loue River, to the east, in the Jura region. If he closed his eyes Monsieur de Villaine could see himself there, standing in the current with his rod, with the music of the birds and the wind and water. He cast his line, his hope, forward. A flick of the wrist and he watched his line soar and then dance down onto the water's surface of brisk currents. He would either catch a prize or, just like that, his line would float back to him, giving him the chance to cast again. In the fly-fishing stream there was no such thing as failure, no family shareholders or critics to disappoint. No pressure.

This was not the case standing in a vineyard contemplating a harvest. Monsieur de Villaine would tell you that every growing season affords the chance for new beginnings, another opportunity to conjure forth from nature and then vinify and bottle some new interpretation of the *terroir.*

Terroir meaning the sum of the natural characteristics unique to each parcel or *climat* of vines: the amount of sunlight and rain an area receives, the pitch and composition of its earth, and, of course, the vines. Roots pull the energy from the earth below, while the leaves harness heaven's sun and draw the rising sap. All of this together, the essence of *terroir,* the very essence of Burgundian winemaking. Although the French Impressionists did not think in such terms, what their very best paintings capture is the magic of *terroir.*

This idea that each vineyard, and then even each *climat* within each vineyard, is its own spiritually charged ecosystem wherein everything is connected in unique alchemy—the grapes merely a manifestation, a by-product of this divine collaboration—is a philosophy that skeptical outsiders have oft dismissed as nothing more than a marketing ploy or misguided pretentious hooey of the French. For the Grand Monsieur the mysterious power of *terroir* was as real as the Savior's death and Easter rising.

In every harvest there was the chance, too, for the vigneron to be born anew, to catch a prize, to achieve poetry and forget, if only for a short time, past missteps, lost loves. Another opportunity to produce a wine more interesting, more pure, than the previous vintage. A chance, if necessary then, for the vigneron to achieve... validation, redemption, to rise again, or, perhaps, bottle what might be his final mark.

There was also, of course, the possibility of crushing disappointment, to fall short of fully harnessing the potential God had provided. Like those vintages the Grand Monsieur had bottled in the 1970s, when he was just starting. Many of those wines were technically correct—"drinkable," as the French say when they are being polite about wine that is subpar—but some were remarkably unremarkable.

In theory, Burgundian winemaking is very simple. The vigneron's greatest challenge is to do as little as possible, to get out of the way of the metaphysical, leaving the *terroir* to nurture and birth the fruit. The vigneron is merely akin to the midwife who facilitates the delivery.

Then comes the pressing of the fruit, where, again, the goal is to meddle as little as possible. Yet the Burgundian process done right must be synchronized to the rhythms of the moon and relies on the soul of the vigneron. At once it is all so simple, and yet maddeningly unpredictable and complex. Like love. Like poetry. Like philosophy.

As a young man, that was all Monsieur de Villaine aspired to do—to travel, to fall in love, to read and write poetry—to study and attempt to unlock the wisdom of the great thinkers. Farming vines, the young Aubert de Villaine thought, he would leave that to others.

As he walked through Romanée-St.-Vivant, Monsieur de Villaine paused, gazed in one direction. Then he took a few more steps, paused, and looked in another direction. He removed his hat and scratched his bald head. In the center, there was a pink spot rubbed raw from so much thinking.

Even the professional meteorologists regard forecasting the late summer weather in Burgundy as a fool's errand. Hail. Rain. Sun. Warm breezes. No breeze. Gentle. Violent. One minute, there is peace; the sun warmly kisses the vines. The next, those storybook clouds turn dark and spit hail that tears through the leaves and pelts and pulverizes the grapes, destroying a whole crop.

In the days to come of this 2010 vintage, Monsieur de Villaine would write in his vineyard journal:

> *At the approach of harvest which we anticipated would begin September 20, it was hard to be optimistic. The weather remained uncertain, governed by west and south winds that brought recurrent humid heat, alternating with rainstorms. We were in the classic situation of the northern vineyards, when often at the end of the vegetative cycle, weather conditions install a well-known scenario: as the warm southern winds furnish the finishing touches to the maturation of the grapes, this heat is also the source of storms that favor the growth of botrytis.*

The maturation had not been uniform. The June flowering—the *floraison*—which had filled the air with that sweet, familiar aroma that ever since he was a child he had likened to the scent of honey, had occurred unevenly throughout the vineyard. The fruit on some vines was further along than the fruit on some other vines. Were the least mature grapes mature enough?

Interestingly, in his vineyard journal, the Grand Monsieur made no mention of the evil that had occurred in his most prized vineyard.

Because the Grand Monsieur consistently produced the greatest wine in the world, everyone who knew anything about wine—and the many who pretended to know about wine—rightly considered him the greatest vigneron in the world. Some went so far as to liken him to a Buddha, to a shaman. For only a spiritual teacher, so went the thinking, could summon from the *terroir* such divinely balanced wines. In fact, it was about that time in the fall of 2010 that representatives of the wine magazine *Decanter* had informed Monsieur de Villaine's representatives in the United States that the magazine wanted to put him on the cover as "Man of the Year."

His advisers at the Domaine's esteemed exclusive U.S. distributor, Wilson Daniels, urged him to seize the public relations opportunity. One of the firm's owners, Jack Daniels, was advising him to go for it. Daniels had nothing at all to do with Jack Daniel's, the famous American whiskey, yet he had a knack for pouring shots of the kind of American straight talk that Monsieur de Villaine had come to value.

The Grand Monsieur would never forget how Jack had stood by him when the Domaine's wines were of such poor quality that they had to be quietly destroyed behind the Wilson Daniels headquarters in St. Helena, California. Jack had stuck by his side, too, when the family tensions threatened to tear apart the Domaine's reputation and even the Domaine itself.

Still, the Grand Monsieur wasn't sure about this award and magazine cover business. Grateful as he was, he didn't want another award. The idea of posing for a cover photograph struck him as immodest and contrary to the Burgundian way. Hoping to entice him, the magazine's people had pointed out that he would be the first Burgundian to ever receive the honor. Though at that moment he remained undecided, the idea that the *Decanter* exposure would be an opportunity for Burgundy to be honored and recognized so publicly dovetailed nicely with his "World Heritage" campaign.

Since 2008, Monsieur de Villaine had been leading an effort to have the United Nations add the Côte d'Or to its list of protected and cherished international landmarks. The list of some nine hundred sites included wonders such as Australia's Great Barrier Reef, the Athenian Acropolis in Greece, America's Yellowstone National Park, and a select few wine-growing regions, such as Hungary's Tokaj area and the Jurisdiction of Saint-Émilion in Bordeaux.

The Grand Monsieur believed the Côte met several of the criteria for World Heritage status, such as being an "exceptional testimony to a cultural tradition" and an "outstanding example of a traditional human settlement and land use." And how could anyone deny that the Côte d'Or "contained superlative natural phenomena or areas of exceptional natural beauty and aesthetic importance"?

It wasn't just the wine magazines and the worldwide legions of oenophiles and critics with their allegedly supernatural palates who regarded Monsieur de Villaine as something akin to the Shaman-Pope and Supreme Professor of wine. Most, if not all, of the world's top winemakers held the same opinion. Certainly, all of Burgundy's winemakers, whether they admitted it or lied and denied it, looked to the Domaine with awe and respect.

One of the neighbors to Monsieur de Villaine's Domaine is the Domaine Georges Mugneret-Gibourg, a winery that itself produces some of the region's most highly regarded wines. For the better part of a year and a half, I lived across the road from Domaine Georges Mugneret-Gibourg. It was comanaged by Marie-Andrée Mugneret-Gibourg. With her porcelainlike skin, youthful eyes, and pixie-style hair, the middle-age Marie-Andrée could have passed for a university student.

One afternoon while we visited and talked of the Grand Monsieur she spoke as if she indeed had a schoolgirl crush on him. Covering her mouth as if she were sharing a secret, she said, "Monsieur de Villaine. You know, you are very lucky to spend so much time with him and to learn about wine from him." Then, in a reverential whisper, she told me, "I have only spoken to him once. Learning wine from him, you must realize, this is like learning physics from Einstein."

Another nearby winery was the Domaine Faiveley, one of Burgundy's most esteemed and dominant domaines, with vineyard holdings throughout the Côte d'Or. Domaine Faiveley is owned by François Faiveley, whose family had founded a company that had helped build much of the modern railway system throughout France and much of Europe. Even in his early sixties, François was a formidable ox of a man who appeared as if he could hammer a railcar together and lay a few miles of track himself if the need arose.

Not so long ago, when the world-renowned American wine critic Robert Parker, whose point-rating system dramatically shaped the modern wine market, crossed a line that François felt should not have been crossed, François led a campaign that single-handedly drove the critic out of Burgundy, essentially forever.

When François oversaw his family's winery, his wines took on his personality. They were famous for being "bold" and "masculine." Now that the winery was directly managed by his son, Erwan, Domaine Faiveley's wines had a reputation for being more "ethereal" and having more "finesse."

One evening, François was hosting a small group of wealthy American collectors at his domaine for a dinner, and as he poured a few of his son's prized wines, he told the group that he believed the high point of the evening would come when he poured some wines sent over as a gift from Monsieur de Villaine. Upon hearing the unexpected news, Monsieur Faiveley's guests were unable to restrain themselves. They clapped with delight and began to speculate among themselves which of the Domaine's wines and which vintage they were about to "experience." François benevolently nodded and smiled. "All of us in Burgundy aspire to what the Domaine achieves," he said. "The Domaine is the standard."

While his guests pretended not to be racing to finish their first pour of the 1999 La Tâche and positioning their glasses for a second, François nursed his first glass as if he were alone. I watched him take a sip, then he raised the glass before his eyes. He rolled the stem between his fingers and gazed over his bifocals into the Pinot. It was if he was pouring himself into the glass.

After a few long moments he turned to me with an apologetic expression on his face, as if to convey he was sorry for having drifted away.

In his gruff rumble of a voice François said: "When I drink this, when I drink the Domaine's wines, what makes it special for me is I think of my dear friend, Aubert de Villaine. I see his face and I think of what he has gone through. I know the sacrifices he makes. During that crime against the Domaine, when the police were investigating, and no one knew anything, I never before saw him so distraught. In his face, you could see this was..."

François's voice trailed off. He turned away and with his massive hand brushed a tear from his cheek.

Unlike the contemporary generations of vignerons who jockey for apprenticeships at the Domaine, Monsieur de Villaine had no degrees in oenology or agricultural engineering from celebrated French universities. Suffice it to say, Monsieur de Villaine could tell a lot from a grape just by looking at it, from considering the skin's color and thickness. Tasting, he would tell you, was the truest way to know.

He squatted down between rows of vines where he knew the grapes were the least mature. He moved aside a canopy of leaves. He did this as lovingly as a parent might brush aside locks of hair from a small child's forehead before a good-night kiss.

He plucked off a small bundle of grapes. He held them just so in his cupped hands, and carefully, with his long, slender fingers, he pushed apart the bundle to examine the quality of the interior grapes. He found only a modest and typical amount of moisture. He tugged off a berry, placed it in his mouth. He bit down on it, ever so gently, just enough to release the juice onto his palate, where he could savor it on the back center of his tongue, and deconstruct it, and cross-reference it with his forty years of tasting pre-harvest grapes.

He spit the grape onto his palm for examination. Poked at it. The purple skin mashed with the yellow-orange mush of the insides. The texture was good. The juice was good and sweet.

Romanée-St.-Vivant was ready. If the weather held—and he judged it would—he could give these grapes even a few more days on the vine. For a moment, Monsieur de Villaine felt his hope float like a line cast high above the Loue.

Then he looked in the direction of his most precious vineyard, the most legendary vineyard in the world. It was just on the other side of the dirt road, marked by a tall concrete cross. Suddenly, everything he wished to forget came back: the ransom notes, the surveillance cameras, the midnight sting operation in the cemetery in Chambolle-Musigny—the murdered vines.

He felt uncertain and sick—and overwhelmed by an emotion he seemed incapable of—anger.

He wondered why God had betrayed him.

The Grand Monsieur walked toward the cross.

The Lord works in mysterious ways, he told himself.

Have faith. Believe.

CHAPTER 2

Unthinkable

The chamber the man had built for himself was small and dark, filled with a kind of disquieting energy. The very same things could be said for his mind.

It was a late fall night in 2009, and inside that small, dark space, he began to stir. A barely audible click, then a light—his headlamp.

He had been lying down, not so much resting as he was waiting for nightfall. Now that it was about 1 a.m., just when he was certain the world around him was asleep, he rose and readied himself.

He was short and squat, with a thick neck and a head like a canned ham. He shuffled about as one tends to do in darkened, cramped quarters. He bumped into things. He was groggy. His breathing heavy. Always, there was wine in his blood.

As the man moved, his tiny spotlight moved with him, darting here and there, illuminating his surroundings in flashes: four walls, a couple of center posts, a roof. The framework formed a chamber no larger than eighty square feet. The limbs that served as vertical supports were anchored into a dirt floor. Wall and ceiling unions bound together by rope and L-brackets. Exterior walls and roof made of blue plastic tarps stretched taut. Blue plastic also covered the floor and on top of the plastic, like a flower floating on a mud puddle, a brightly colored doormat. The overall aesthetic of the place was akin to Robinson Crusoe meets the Unabomber.

The interior felt vacuum-sealed. The trapped air was greenhouse humid, weighted atmosphere, invisible cobwebbing, stale. Tolerably uncomfortable. That the space was subterranean, burrowed into the earth like a giant weasel warren, was palpable. So, too, were the smells: plastic of the tarps, dirt, body odor, laundry in need of washing, pungent cheese, stale wine.

Along the east wall was a cot, also made of tree branches and topped with a foam mat and a sleeping bag. Against the west wall a hot plate, pots and pans, and a narrow table—a plywood top affixed to tree-branch legs. On the floor, around the interior perimeter, plastic bins were neatly stacked, even under the cot and table. Tight. Well organized. All in all, an efficient use of meager space, correctly giving the impression that this was someone accustomed to making use of a confined room.

An array of items was scattered on his makeshift table: a clock-radio, an MP3 player, work gloves, a jar of *moutarde*, a Tupperware container of *carottes*, a small wheel of Lepetit brand cheese, a pair of bent and smudged bifocals, a diarylike notebook. And there was a magazine—one of those large-format, richly colored glossies. In the headlamp's light the magazine's cover shined like a polished pearl. It was titled *Bourgogne Aujourd'hui*, or "Burgundy Today," a periodical dedicated to *Les Vins et les Vignobles de Bourgogne*, "The Wines and Vineyards of Burgundy." One of the stories in that issue was a feature on the legendary Domaine de la Romanée-Conti.

On just about any list of the world's twenty-five top-rated wines, the Domaine de la Romanée-Conti regularly places seven: Richebourg, Échézeaux, Grands Échézeaux, La Tâche, Romanée-St.-Vivant, the Domaine's only white *grand cru*, Montrachet, and the world's very best wine, which is the winery's namesake *grand cru*, Romanée-Conti. For its unparalleled and sustained excellence, the Domaine de la Romanée-Conti is known by wine critics and serious oenophiles around the world and frequently referred to by its initials, or simply as the Domaine.

The article noted the insatiable market—the legal and otherwise "gray" market—for the wine. This, despite the fact that not surprisingly the wines also happen to be among the very most expensive in the world. A bottle of the Domaine's least expensive wine, Échézeaux, in the most recent vintage available, which is typically the least expensive vintage of any wine, was then going for about $350. For a single bottle of the Domaine's priciest wine, Romanée-Conti, the cost was roughly ten times that of the Échézeaux, at $3,500.

As astonishing as those retail prices were, they were misleadingly low. Because the Romanée-Conti vineyard is so very small—4.46 acres—and because its yield is kept low, the wine is extraordinarily rare. What's more, the Domaine itself keeps strict control over its allotted sales to distributors and select individual clients who buy up the wine in pre-sale orders before the wine is even bottled.

Frankly, there is almost zero chance of finding a bottle or Romanée-Conti in your local fine wine retailer at all. Thus the shady back-channel "gray" market and the wine's booming Internet and auction sales, where the price for a bottle of the most recent vintage of Romanée-Conti—which for all practical purposes is the baseline price—was then more like $10,000 per bottle.

Bottle for bottle, vintage for vintage, Romanée-Conti is the most coveted, rarest, and thereby the most expensive wine on the planet. At auction, a single bottle of Romanée-Conti from 1945 was then fetching as much as $124,000.

In one of the photos that accompanied the article the Romanée-Conti vineyard indeed appeared to be a remarkably tiny patch of earth at the base of a gently sloping hillside. Nothing at all outwardly different from the ocean of vineyards around it. A low stone wall lined a portion of its borders. On top of the wall stood a tall, concrete cross, its elongated shadow swimming across the leafy canopy tops behind it.

In another picture, a draft horse tugged a plow between the vine rows. It was a contemporary photograph, to be sure, which made the antiquated farming technique appear all the more odd. These pages of the magazine were dog-eared and pen-marked, as if the man had lain in his cot studying the pages over and over again.

Also among the items on the makeshift table were three bottles of wine: a Côtes du Rhône, an Écusson Grand Cidre, and a Hérault. All of them drunk into varying degrees of fill levels. The label on the bottle of Grand Cidre promoted it as *cuvée spéciale*. This distinction, as the man had been formally educated and generally raised to recognize, was little more than one of the wine world's many gimmicks.

There was nothing especially *spécial* or *grand* about the Grand Cidre, or, for that matter, the other two bottles—except maybe that they had been in the special sale section at the local *supermarché*. These wines were what the French referred to as "common." The sort of plonk you'd pick up for a few euros at the local SuperU if you wanted to wash down a microwavable quiche, or, if you were in the market for something to polish off in order to forget, to ease nerves, or, as was now the case for the man, to gin up what might pass for courage before executing the unthinkable.

His selection of wines from the Côtes du Rhône and Hérault regions of France, the man knew, amounted to a perverse irony. It was in the southern part of the Rhône-Hérault region, a century and a half earlier, that a trespasser had crawled into the vineyards and launched an attack on vinestocks that wiped out nearly every vineyard in France. It was a nationwide economic issue, a countrywide identity crisis. Authorities of the time dubbed that menace *Phylloxera vastatrix*—aka the "devastator of vines."

And now here he was.

Over the years, for previous jobs—"projects," as he liked to call them—the man had relied on pipes, handcuffs, guns. During the job he was on before this one he had made a point of laying out all three of those tools, piece by piece, ever so slowly, on the kitchen table of his female victim in order to terrify her into compliance.

On that job, which he executed in another famous French wine region, Bordeaux, the man had proven he would pull a trigger, even if it meant taking aim at *les policiers* of the gendarmerie. However, he had done enough crimes, done enough time, exchanged enough gunfire, to realize there were easier ways to take a buck. This current project with the vines, it was not that kind of job; those kind of tools and that kind of risk were not necessary. That's what the man told himself. Still, he kept a pistol nearby, just in case.

His headlamp beam settled on a container not much larger than a lunch box. It was on the floor near the cot. He opened the case. Inside was a battery-operated drill. A Black & Decker. Not far from the drill, a few syringelike devices similar in size and appearance to turkey basters. He grasped one of the syringes—his fingers were as stubby as hors d'oeuvres sausages—and reached for a plastic gallon container and from it clumsily poured a liquid into the syringe.

His heavy breathing became more strained as he pulled on calf-high green rubber boots. From a hanger dangling on one of the crossbar tree limbs he removed a long hooded rain jacket. Green and rubbery like the boots, it wasn't so much a coat as it was a hooded cape. He put it on, tucked the drill and syringe into a pouch belted about his waist, and turned to the door.

The hatch, too, was made of sticks. He pulled on the door, once, then again. The bottom of the door, as always happened, had snagged on the dirt ground. He opened it just enough to squeeze through.

Outside, the chilly air sent a shiver up his sweaty back. He scrambled a few feet up into a small clearing surrounded by dense woods. The night sky was as black and as soft as tuxedo satin. So many stars. The moon was full and bright. Liquidy, as if the orb were filled with white lava. Wisps of clouds crossed its face. There was no need for the headlamp. He clicked it off. Doing so decreased the already slim chance of his being noticed.

He waited a moment to give his eyes time to adjust.

Sometimes, at about this hour, there were the sounds of wild boar cracking through the woods around him. Off in the distance, straight out in front of him, to the east, he could hear the faint whooshing whistle-groan of the TGV. The high-speed train streaked along tracks either bound for the city of Dijon in the north or heading south toward Beaune.

The train was how he would make his getaway. He was so

close. He just needed to finish this last critical bit, then collect the money, and take his cut, and be gone.

As he stood there above the shelter, it would have been understandable if the man felt a sense of accomplishment. Viewed from this perspective his handiwork was all the more impressive. His flat, square box of a cabin was inside a square ditch. The walls, which were about six feet high, were almost entirely below ground level. The exterior was wrapped in olive-colored plastic tarp. The roof, covered over with leaves and twigs, was indistinguishable from the forest floor.

Some of the most skilled detectives of the French national police soon would come to learn you could fly a helicopter over it a dozen times and not see it. Hell, you could be standing right next to it and never realize it was there. Investigators would marvel at the structure. The excavation alone, not to mention everything else involved in erecting and equipping the place—sturdy, water resistant, bivouacked into the earth, buffered from the wind, masterfully camouflaged... it had taken months.

The man headed off into the woods.

Within minutes he emerged from the forest and stepped into a panorama that was as expansive and as ethereal as his shelter was small and squalid. A silhouette in the hooded cape, he stood atop a hill, his pulse throbbing within his thick neck. As he had done so many nights before, he scanned the landscape to make sure all was clear.

In the moon's glow, the view was empowering; the world was at his feet: Spilling down the hillside and then everywhere was a vast patchwork of vineyards. Sprawling straight out in front of him, to the east, and to the north and south, seemingly without end. Row after row they unfurled, barely separated from one another by ribbons of fallow land or narrow road. The vines were frost dusted and barren, twisted and vulnerable, like the skeletons of arthritic hands reaching for spring.

Just as he had come to expect, just as it had gone on the previous nights, no one else was out. The only movement was the headlights out east, well beyond the vines. The cars traveled on Route Nationale 74. Beyond the RN-74, the train tracks. He could once again have his way without fear of detection. It never ceased to amaze him, to please him, that so much value was just left there unprotected.

The hill—the *côte*—on which he stood is part of a formation that stretches through much of the Côte d'Or, some twenty miles to the north and twenty miles to the south. He turned right and took a footpath south.

With the vines to his left and the tree line on his immediate right, he took the path for about a half mile. He then descended the slope and entered the vines.

The vine rows continued as the hill flattened out and then right up to the edge of the small hamlet, less than a mile away. The tiny town's skyline was humbly marked by a church steeple. Walking through the vines in the direction of the town, he exuded the purpose of someone who knew precisely where he was headed and what must be done when he arrived.

Midway between the hilltop and the town, on the upper edge of a vineyard that was at the base of the gently sloping hillside, he stopped and fell to his knees. Had anyone happened upon him he might have appeared to be praying. Which he knew would not have been unusual.

For months, he had been casing the vineyards, on bike and on foot. He watched as people from all over the world arrived every day at that vineyard. Some were your typical tourists. Many, however, were zealots, passionate about Burgundy wines. Like pilgrims traveling to Mecca, these "Burghounds" came not so much to see the vineyard, but rather to behold its presence. Often these pilgrims quite literally would kneel. Always they would go to the tall, concrete cross towering over the vines and snap a photograph.

Affixed to the low stone wall, not far from the cross, was a sign. Words written in French and in English stated:

> MANY PEOPLE COME TO VISIT THIS SITE AND WE UNDERSTAND. WE ASK YOU NEVERTHELESS TO REMAIN ON THE ROAD AND REQUEST THAT UNDER NO CONDITION YOU ENTER THE VINEYARD. THANK YOU FOR YOUR COMPREHENSION. —THE MANAGEMENT

Truth be told—and the Management realized this—it was not unusual for a visitor to dismiss the sign; to throw a leg over the wall—wait for a moment as if they half expected an alarm to sound—then throw the other leg over the wall and timidly scurry a few feet into the vines and pluck one of the grapes for a taste, or to grab a handful of soil, or even to pocket one of the small chunks of white stone peppered throughout the vineyard.

It was with a mix of pride and benevolence that the Management had resigned itself to the reality of these occasional acts. Not that the Management encouraged such behavior or would ever look the other way if they were present to witness such an intrusion, but they realized these lawbreakers do what they do only out of admiration, adoration even; they meant no harm; they were misguided but well-intentioned. They were like the tourists who ignore the many clearly posted signs at the entrance of the Sistine Chapel and nevertheless snap photographs of Michelangelo's ceiling masterpiece.

Only this vineyard was more ancient; its history every bit as epic, and, to many, even more sacred than that of any of Michelangelo's sixteenth-century paintings. Unlike a masterwork painting, this scene didn't seem to come alive—it was alive. And while the wine it produced was out of financial reach for most mortals, locked away in cellars of wealthy collectors, as far as the vineyard goes there were no alarms, no security personnel, no cameras—the vineyard was right there in the open, just off to the side of a strip of crumbling road, within reach of everyone, vulnerable to anyone.

The man got down on all fours. His barely moonlit face hovered inches above where the vinestocks were married to the earth. The tendrils of his hot breath rose into the night. The topsoil was cold and hard, but scratch just beneath the surface, dig down a few inches as the man did and there was...

Mon dieu, le senteur.

Nutrient-rich, rocky soil that had been churned over and over again thousands of times, hundreds of thousands of times, so that the earth could breathe and the vines could drink, hydrating roots that at that very moment, every moment, pushed through, around the rocky geological layers below—pushing through both because of and despite nature.

Le senteur.

It filled his nostrils, cut to those parts of his brain that triggered memories of his childhood.

His father.

His earth.

His vines.

Here, though, the smell was different.

This earth emanated a musk. A musk infused with the scents of salty ocean, minced seashells, a wet minerality—like chalky stone damp with spring rain.

Here the geology was luscious. This earthiness, odd as it may sound, was mouthwatering. There was an aromatic come-hither temptation to taste the dirt, to want a "droplet" of its textures to roll and spread, and rest in the back of the mouth. A musk that caused the tongue to fatten with anticipation of...a sip.

He produced the cordless drill and the syringelike device. He pressed the drill bit into the vinestock, just where the vine disappeared into the earth, and he began to drill. Into the *pied de vigne*—the foot of the vine.

The sound, the soft whir of the drill's motor, registered as nothing in the vast quiet. In the distance, the quaint town, with its shutters drawn, was too far off, too asleep, too trusting to notice. No one in all of Burgundy—really, no one in all the world—had ever contemplated that anyone would conceive, let alone execute, such an act, such a sacrilege.

Crouched among the vines, the man shifted his attention to a neighboring vinestock. It was less than a yard away from the one he'd already drilled. With the Black & Decker, he repeated the same procedure on the foot of that vine.

Next he took the syringe, inserted it into one the holes he had drilled, and injected some of the syringe's contents. He did the same to the other vine, emptying out the rest of the liquid. From his pouch, he fished out two tiny wooden plugs; he pushed one into each of the holes he drilled and returned the soil around the vinestocks, best as he could, to the way he found it. As if they had never been disturbed.

The man understood perfectly what he was doing in terms of the crime, in terms of the science of the vine—the viticulture. He could grasp the localized smallness and he understood destruction. The implications of his actions, the transcendent largeness of it, that was something he could not comprehend. For him, this was about the money. Well, if he had been forced to admit it, it may also have been about a personal vendetta.

Matter-of-factly, he collected his equipment and made his way up the hill. He emerged from the vines, traveled the brim of the *côte*, and again vanished into the dense tree line.

Inside his underworld studio he hung up his hooded cape on the hanger and poured himself a glass of the *supermarché* swill. A toast to the final stages. The two vines he had drilled were among the more than *seven hundred* vines that had been drilled in the vineyard of Romanée-Conti.

He lifted his MP3 player from the table and pushed the earbuds into his meaty head. Mozart, as the police would learn from the statement of someone else involved, was the man's favorite. The music poured into him, flowed through him. The man knew that come spring, the sap travels through a vinestock, carrying nutrients to the outer extremities, infusing the precious fruit. Similarly, the music traveled through him.

According to the reams of information that would be gathered by investigators, viticulturists, and scientists, then photocopied, stapled, scanned, shared with the head of the Police Nationale in Paris and the courts, and then finally filed away in confidential dossiers, where it was hoped the unprecedented case would quietly disappear as if it never happened, when this project on the vines was over, when the money was divvied up and the man had his cut, his dream was to buy an old church with an organ. His dream was to learn to play Mozart on the organ, which was how he believed Mozart was meant to be played.

CHAPTER 3

Conti

As he stepped into a Paris night late in the summer of 1755, Louis-François de Bourbon presumed he was under surveillance. He figured spies had eyes on him that very moment. He had no doubt they had been intercepting and inspecting his mail. Someone had been clumsy about removing the seals on his letters. Melting away wax marques by candle heat and then replicating and reapplying counterfeit seals was an art that required surgical attention to detail. It was a task that needed to be assigned to the steady hands of a master; whoever had been slicing into his correspondence was no master.

Louis-François was an expert on such matters. That was why he rarely committed compromising words to ink. Instead, if he had to write such a note, he did so only by pinpricks. Furthermore, he took steps to ensure the recipient understood in advance to immediately destroy the correspondence after reading. Having noticed that his mail had been breached gave Louis-François a counterintelligence advantage: the opportunity to disseminate disinformation to throw them off his trail. He was quite good at that sort of thing, too.

No matter, he was confident they had no idea of his exact plans. Despite what some at Versailles thought of him, the prince was not so full of himself to believe he was infallible. The prudent course of action was to move under the cover of night and to not underestimate his adversaries. And so he considered the possibility of operatives lurking nearby.

Espionage was a game Louis-François played better than anyone. Really, it was his game. He was the French spymaster, by virtue of practice and by occupation. There would have been no Secret du Roi without him. He was the architect of that spy network of mid-eighteenth-century France. He was the one who oversaw the Secret du Roi's recruiting and managing of the Crown's agents throughout Europe.

There weren't many, if any, tactics Louis-François had not seen or employed. Now that he himself was the subject of intense surveillance it concerned him, of course, but it most certainly did not unnerve him. Part of him found the irony of it, the personal challenge—and indeed he viewed it as a personal challenge—rather delicious. He carried on just as he would have advised his own agents to do: cautiously, but with typical Parisian, aristocratic composure, as if nothing at all were out of the norm.

Which was not the case. Moments that set in motion seismic historical events, that compel men to take up arms and kill, that change the balance of world power, that overthrow kings—this, he thought, was one of those. He had to believe that. Or else, what point was there in all of the risk?

It was August, the month of his forty-first birthday, and Monsieur Louis-François de Bourbon, the Prince de Conti—a royal-blood cousin of King Louis XV, and also His Majesty's de facto chief of staff—climbed into a carriage bound for a clandestine rendezvous that by definition of the king's law constituted conspiracy to commit the highest treason.

More than anyone else it was the Prince de Conti who had the ear of King Louis XV. Their relationship was a subject of great interest within the corrupt and catty royal court. The nobles and their servants whispered about it, and noted it in their memoirs and correspondence thusly: "People are always astonished by the intervention of the Prince de Conti in affairs of the state." His "intimacy" with the king, his access to His Majesty, and influence upon him are "quite remarkable." The Prince de Conti, alone, would daily enter the king's private study "by the backdoor carrying great portfolios." Often not emerging until hours later.

There was no doubt the two men talked strategy for France's foreign affairs, which were rapidly escalating into military conflicts. France's claims in North America were being challenged. The previous spring, in 1754, over in America, a local British militia under the control of a Lieutenant Colonel George Washington had ambushed a contingent of French forces.

It was one in a steady stream of ongoing guerrilla battles between the two countries in that foreign land. This one, though, occurred in the critical Ohio Territory and became a spark for what was now, more than a year later, all-out war. This "French-Indian" battle exacerbated tensions between the two powers, already fighting over shipping routes; it contributed to their taking opposing sides in a war between Prussia and Austria, in which Russia and Spain were also invested. All of it one big, bloody international mess that was turning into a Seven Years War.

A testament to the truly top-secret nature of the meetings between the prince and the king, no one at the court had any gossip about the specific discussions of their meetings. Due to the mystery shrouding their sessions, as one member of the royal court wrote, "people had difficulty understanding what can be the nature of their work." That is not to say that Conti was someone who otherwise kept a low profile.

He distinguished himself as a character among characters. A prerevolutionary James Bond. The prince left such an impression on Madame de Genlis, a noted contemporary writer and noblewoman from Burgundy, that she mused on him in her diary:

"The Monsieur le Prince de Conti was the only prince of the blood who had a taste for the sciences and for literature, and who knew how to speak well in public. He was strikingly handsome, with an imposing figure and manners. No one was able to pay a compliment with more finesse and graciousness and, despite his success with women, it was impossible to discern in him the slightest nuance of fatuity. He was the most magnificent of our princes."

Born on August 13, 1717, into a family with Burgundian roots, and one of the most noble of France's families, Louis-François studied philosophy and the arts, having a particular fondness for Mozart. Most notably, perhaps, he was a lover of love, and, as was common for the noblemen of the times, more often than not with women other than his wife. Louis-François had been fourteen years old when he wed his cousin, the fifteen-year-old Louise-Diane d'Orléans, the youngest daughter of the duc d'Orléans, Philippe II.

Louis-François had married into quite a family. When King Louis XIV died in 1715, he had already buried his son and the grandson who would have been next in line for the throne. The monarchy, then, had to wait for his great-grandson, Louis XV, who was only five years old at the time of his grandfather's death.

Until Louis XV was old enough to wear the responsibilities that came with the crown, the duc d'Orléans, Philippe II, served as the Regent of the Kingdom. The union of his daughter, Diane, and Louis-François was celebrated in grand fashion at Versailles and, of course, had been arranged for purposes of bloodline politics. The marriage did nothing to discourage the teenage Louis-François from promptly beginning an affair with a mistress inherited from his uncle. (Louis-François's uncle was moving on to a Parisian dancer.)

The Prince de Conti was the type of renaissance man who continued to engage in picaresque, libidinous adventures, relishing every opportunity to insert himself into affairs of all sorts. Along with the women, there was the wine. At parties, whether at his cousin-king's palace of Versailles, or at one of his own residences—the Palais du Temple, where he had a rank among the Knights Templar, or at his private residence, Hôtel de Conti in Paris, or at his retreat at the L'Isle-Adam, a couple of hours' carriage ride south of the city—no soirée was complete without the prince filling a beautiful woman's ear with charm and her glass with exquisite wine.

Not long after the covert business of that summer in 1755, the "inviolable secret," as Conti himself had begun referring to it, would reach its stunning, almost inexplicable dénouement: The prince would commission a painter to memorialize one of his own dinner parties.

In the scene that Michel-Barthélemy Ollivier would paint, a dozen white-wigged nobles sit around a long table, amid the warm glow of candlelight. In a nearby corner, a harpist strums. In the room so vividly alive with the buzz of intimate conversation and cascading string music, Conti looks into the eyes of a woman on his right, his mistress du jour, while his left hand seductively caresses the neck of a bottle of his private reserve, which was then known as La Romanée.

By the time the prince would acquire the Burgundian vineyard its Pinot Noir would already have a reputation for being sensationally smooth, stunningly complex, the perfect balance of seductive and powerful—much like Conti himself. However, legend would have it there were other reasons the prince would go to the great lengths he would to acquire the vineyard—also involving surreptitious maneuvering. Reasons that were only now in his present secret matter beginning to take shape. Before there would be Burgundy, there would be Paris, and if the prince had his way, there would be revolution.

That Conti was so openly dashing yet so politically discreet was one of his many dichotomies. The image of the bespoke, silver-tongued playboy belied the prince in full. He was a decorated war hero several times over, a murderer, a spy—a double, maybe even a triple, agent. He was a fiercely intelligent operator, and generally speaking, an illusive chameleon.

One of the prince's fellow noblemen astutely sized him up as "a composite of twenty or thirty men. He is proud, he is affable, ambitious and a philosopher, at the same time; rebel, gourmand, lazy, noble, debauched, the idol and example of good company, not liking bad company except by a spirit of libertinage, but caught up in much self-love."

Considering the prince's shrewdness, he may have sustained such a colorful and charismatic dandy-man persona to distract from his covert and most grave sleights of hand. A misdirection by façade. By that August of 1755, Conti was someone whom his cousin-king and Louis XV's omnipotent mistress, the Madame de Pompadour, had come to fear and mistrust. The king and Pompadour, Conti had no doubt, were the ones who had ordered the postmaster to intercept his mail. They had put him under the surveillance of the French police.

The mission was overseen by Lieutenant Nicholas-René Berryer and a contract agent, Soulier de Puechmaille, aka Lagarde. Lagarde had been recommended for the task by none other than the archbishop of Avignon. The Crown, the church, just about everyone benefiting from the monarchy's stranglehold on the people, considered Conti a threat to all that was royal and holy. Their suspicions were warranted.

That night, as the prince made his way to his clandestine meeting, if one of the spies would have found a way to casually emerge from the shadows and inquire the prince about his destination, Conti might have offered an explanation that he was en route to conduct official business of the king, a mission for the good of France. For it was exactly the sort of politically deft response for which Conti had such a gift: a shred of fact that provided just enough cover for the whole treasonous truth.

Rattling over the cobblestones, navigating Paris's narrow *rues*, the carriage almost could not avoid jolting to starts and jerking to stops, twisting with expected unpredictability into the abrupt turns of the capital city. It would have been prudent of Conti to instruct his driver to make a few unnecessary turns along the way to make the route all the more circuitous and harder to follow. During the day, the urban labyrinth teemed with the activity; a mosh pit of nobles and peasants, where it was difficult to discern vice from virtue. In the words of a writer of the time, the city was "a rapid and noisy whirlwind."

With a population approaching 25 million, France was three times the size of its mighty and increasingly nervous neighbor

England. Nobles and clergymen together—the First and Second Estates—formed the 2 percent of the population that controlled most of the country's wealth. The poorest of everyone else, the Third Estate, labored to buy the bread they could already barely afford. Peasants had petitioned their aristocratic landowners to invest in agricultural improvements or, at least, to tax them less so that they themselves could modernize and more efficiently harvest grains and wheat, thus producing more bread and making it more affordable. Such requests had been met with indifference.

With the Catholic Church's blessing, nobles openly scoffed at labor as something only bourgeoisie did, in order to earn the taxes aristocrats could "invest" in the church and their own leisurely pursuits: patronizing the arts, which were often odes to themselves or packed with messages to reinforce the necessity of classism; and building their grand palaces, like Versailles, where the Prince de Conti himself kept an apartment; and throwing decadent parties. In the tradition of the late King Louis XIV, every nobleman worth his unearned *livres* peacocked on the dance floor. Small fortunes were spent trying to outdo the Italians in the latest fashions.

Among the masses squeezed into the poorly defined Parisian city limits, wigged noblemen wore splendid collar-band waistcoats and polished, buckled, high-heel shoes; the powdered noblewomen were tightly corseted inside brightly colored hooped skirts of the finest imported fabrics, and many of them wore their hair styled in a towerlike fashion—the gravity-defying pompadour style made popular by the madame above all madames, Madame de Pompadour. Aristocrats promenaded on their way to doing positively nothing at all, doing their best to gracefully pass untouched through the masses.

The *petites gens*—the small people: workers, servants, artisans, shopkeepers—hurried about, fortunate to have jobs. Pickpockets, whores, and beggars in their tattered clothes, often infested with lice, ill and in some cases deformed by disease, assertively targeted their marks. Many nobles traveled in decorative, phone-booth-like sedans carried by servants or in carriages that rolled through the streets with footmen jumping from the carriage rails to shoo off the glut of commoners to make way.

As the rich and poor rubbed against one another, economic and religious friction sparked tensions that the media had lately fanned into flames. Reading had become more than the fad that French aristocracy thought it would be, or rather hoped it would be. For reading meant education and thought, and thereby enlightened challenges to the status quo. In the cafés and salons literate members of the Third Estate drank the "common" wine made from the Gamay grape, and read the papers and pamphlets and works by the likes of Jean-Jacques Rousseau.

Only a year earlier, in 1754, Rousseau had published his essay "What Is the Origin of Inequality Among Men? And Is It Authorized by Natural Law?" In it Rousseau wrote what those who had no voice longed to have heard:

> Such was, or may well have been, the origin of society and law, which bound new fetters on the poor, and gave new powers to the rich; which irretrievably destroyed natural liberty, eternally fixed the law of property and inequality, converted clever usurpation into unalterable right, and, for the advantage of a few ambitious individuals, subjected all mankind to perpetual labor, slavery, and wretchedness.

It wasn't just economic oppression—and oppression was now the word—it was also religious oppression. It was the Catholic state's oppression of Protestants. On his deathbed, Louis XV's predecessor had reaffirmed that Catholicism was the only religion in France. All subjects must kneel before Christ or else be regarded as traitors, and, as was the case for some Protestant pastors, be put to death. Protestant churches routinely were burned to the ground.

Rendered pariahs, the country's community of Protestants peppered about the country gathered in open secret to worship and rallied one another in their treasonous conviction that they had a right to worship as they saw fit. Meanwhile, priests steadfastly preached the divine right of kings, now the authority of King Louis XV, who not surprisingly held the view that the Huguenots were a lesser species in need of conversion. Louis XV issued laws reinforcing Catholic hegemony, reiterating that only Catholic births, marriages, and deaths were legitimate—legislative genocide.

The Huguenots were not the only religious group feeling persecuted. Nudged by the church, Louis XV launched a political crusade against a Catholic splinter group. The Jansenists believed that God alone, through his mysterious and divine ways, determined who had grace, and that no man, no clergyman, not even the king himself could determine who was forgiven in the eyes of the Lord. As far as the establishment was concerned this was both religious and political heresy. Louis XV supported his bishops who banned Jansenists from receiving Catholic sacraments.

Protestant and Jansenist leaders appealed to their king and more directly to the magistrates of the French law courts, the *parlements*, and in particular, to the most influential of all the law courts, the Parlement of Paris, for equality, or at least some compromise that would allow them to live free, as *citoyens*.

Parlementary magistrates wanted to provide a degree of what amounted to civil rights to accommodate the religious sects. To ease the tensions, the *procureur général* to the Parlement of Paris, Guillaume-François Joly de Fleury, suggested that the king consider recognizing Protestant marriages as civil unions. The reality denied by the monarchy was that Catholics lived quite nicely among Huguenots and Jansenists. Not only that, the Protestants were integral to the day-to-day French economy.

If the king would not reconsider his views for altruistic purposes or for reasons of economic necessity, the magistrates pointed out, there was the growing fear that the disenfranchised groups would coalesce into an uprising. Bloody riots already flared around the country. Passions were especially volatile in the south of France, where Protestant leaders like Crown-defined enemies of the state Pastor Paul Rabaut and Jean-Louis Gibert preached with rhetoric that was becoming more and more militant. Rabaut was of the mind that "the persecution is becoming stronger from day to day; and for quite a while, we have had so many reasons to cry, Lord, save us, for we are perishing." Gibert was brazenly defiant. He proclaimed that his flock was prepared to "break the bonds of our captivity and uphold our liberty and that of our religion, even at the cost of our lives."

Outside the echo chamber of the royal court, the reality had become so intense that in that spring of 1755 the Parlement of Paris refused to ratify Louis XV's decision to allow the church to ban Jansenists from receiving the sacraments. Taking the position that the king's policies required the approval of the *parlement*, the magistrates simply packed up and went on strike. In shutting down government business, the magistrates thumbed their noses at the king, also tabled approving his funds, putting a crimp in his debauchery. Louis XV dispatched musketeers to arrest four of the most vocal opposing magistrates, and he sent two hundred magistrates into exile.

The more steps Louis XV took to centralize his power, the more it fractured. While the French military was at war in several international theaters, by that summer of 1755 he feared a domestic revolt. Louis XV began to hear the protesting voices in his head like a relentless monastic chant, the vibrations of which began to shake his throne and his mental stability.

The king saw traitors where there were none, and trusted aides were there were traitors. As Pompadour's personal handmaid wrote in her diary, the king had long been "habitually melancholy"; now he began to sense threats from all directions. While some members of the court whispered about paranoia, death was indeed coming for the king. Perhaps at that very moment it was choosing its weapon and path. A plot for assassination was under way.

In his private study, just as he had on matters of foreign affairs, the king turned to the man who had been his friend since they were children; the one person he trusted and respected more than any other; a man who was equally trusted and respected by the *parlement*, by the Protestants and Jansenists, by the French military, and for that matter, by the French people—his cousin, seven years his junior, the Prince de Conti.

During the day, when Conti's carriage would travel through Paris, pushing through the crowds, his street-level perspective afforded him an intimate view of the volatility of the times. Everyone, everything, it was all right there in the streets of Paris: the people together before him, around him, so tightly mashed together, yet divided. Such that the country maybe could not stand. The inequity, the resentment, the hate: All of it was seething. He could smell it as plainly as he could smell the raw sewage dumped into the streets. The operative in Louis-François knew that such dissension could be a valuable tool. It could be harnessed; it was a power. He carried these observations with him into the darkness and his secret rendezvous.

The secret meeting had been facilitated by Conti's aide, Nicolas Monin. Monin had served in the army under Conti and remained by his side when the prince returned to the royal court in the mid-1740s. It was as King Louis XV's trusted chief of staff that Conti had been empowered to pursue delicate diplomatic missions and persuaded the king to agree that a network of spies was necessary to gather and relay intelligence throughout Europe via codes and other means. Monin had been an integral part of helping Conti build that infrastructure and managing reconnaissance assignments.

Recognizing the extraordinary nature of what was to be discussed that summer evening, Monin arranged for the treasonous appointment to take place down on the waterfront, in an abandoned building on one of the anonymous quays that wind along the banks of the Seine. Inside Conti greeted his visitor, none other than the Protestant pastor and wanted enemy of the state, Paul Rabaut.

It was the second meeting for the two, the first having occurred just a few weeks earlier. There was less of a need for small talk before getting to business. It would have been typically gracious of the prince to thank Rabaut for once again making the long trip from Nîmes to Paris. Rabaut, a man of devout faith and with immense respect for the prince, was always expressing his gratitude to Conti for his continued interest in the Protestant cause.

Because the meeting occurred around the prince's August birthday, Conti would have had his age and mortality on his mind. Considering the endeavor in which the two of them were engaged, it was reasonable for Conti to wonder if he would live to see his next birthday.

Rabaut had contacted Conti months earlier, at first writing to him care of intermediaries, and then, having been assured that the prince sympathized with the Huguenots, to the prince directly. He had asked the prince if he would lobby the king to reconsider his policies regarding the Protestants. The prince had agreed.

The ostensible reason for the secret discussion now was a status report. The prince shared with Rabaut whatever progress he was making in his private sessions with the king. Rabaut briefed the prince on the state of affairs of the Protestants down south. In short, none of it was going very well. King Louis was unwilling to budge in any meaningful way and the Huguenots only grew more restless.

Before long, Conti began ever so softly exploring Rabaut's interest in an armed uprising. The prince wanted to know just how united Rabaut's parishioners were. He asked if they had access to arms. They did. He asked if they would they be willing to use them. Rabaut suspected they would. The prince wondered how often Rabaut or any of his colleagues communicated with their Huguenot brethren in England. More specifically, the prince was keenly interested in whether the Protestant leadership had any communication with the British military or government.

Indeed, the Protestants were communicating with England. It was likely that Conti already had some knowledge of this, as he had his own well-established lines of communication to London courtesy of the Secret du Roi network. The prince also had an idea, which he was now softly floating to his Protestant contacts like Rabaut: a nationwide Protestant uprising triggered by an English invasion on the southwestern coast of France.

Agents involved on both sides of the English Channel had begun to call the plan the "Secret Expedition."

CHAPTER 4

Edmond's Hope

The thing about fate is that you never see it coming.

Of course, you can plan and work toward a goal, but that all-encompassing raison d'être, that's another story. Whether it turns out that it is all God's plan, or, as the French philosopher Jean-Paul Sartre put it, that "we are all condemned to be free," Sartre was right about this much: "existence precedes essence." It's only after one has lived and discovered destiny, or maybe surrendered to it, that the obviously significant moments of the inevitable trajectory can be appreciated. But no small boy has such thoughts or ponders his calling.

So on a spring morning in 1947, when he accompanied his grandfather on one of his weekly visits to the family's Domaine, eight-year-old Aubert de Villaine was merely along for the ride.

Edmond Gaudin de Villaine was scheduled to arrive at the Domaine in the village of Vosne at his usual time of 10 a.m. The drive from Moulins, the village where they lived, typically took two hours, but often they traveled to Vosne the night before. On that day—one of the very few days Aubert would ever make the trip with his *grand-père*—they had arrived a few minutes early.

For Edmond, a former military man, punctuality was a necessary form of respect. He did not want to prematurely disturb Monsieur Louis Clin and the Madame Geneviève Clin, who lived at the Domaine, managed the winery's day-to-day affairs, and, first and foremost, tended the vines. When little Aubert and Edmond arrived at No. 1 Rue Derrière le Four—tucked in the bend of the cobblestone street, shaped like a boomerang and no wider than an alley—they stood at the winery's large red iron gates until Edmond's watch showed precisely the hour. Only then did he press the buzzer.

Madame Clin greeted them. The wife of a retired army officer, she was as fiercely particular as any commanding officer her husband had ever known. Everyone in Vosne knew not to test Madame Clin's commitment as the Domaine's sentry. Each day she would walk the vineyards, inspecting the border stones of the Domaine's holdings, and God help any neighbor she caught "mistakenly" moving any of those stones and encroaching on the Domaine's sacred soil.

"*Bonjour, Madame Clin,*" Edmond said. "*Comment vas-tu?*"

"*Bonjour, Monsieur de Villaine. Très bien. Très bien.*"

She ran her hands along the front of her skirt, smoothing out nonexistent wrinkles.

Although they had known each other for years, Edmond and Madame Clin addressed one another with a warm formality informed by Old World French etiquette and great mutual respect. Edmond valued the Clins' service and unwavering dedication to the Domaine. As tough as Madame Clin was, she had a wink in her and softened around Edmond. She knew what he had endured and she understood what the Domaine meant to him. It was a legacy that had come to him by way of a great love and loss.

In 1906, Edmond had married Marie-Dominique-Madeleine Chambon, the woman he thought would be the only love of his life. Marie-Dominique was from a family of wealth and prominence, a status derived from her great-grandfather Jacques-Marie Duvault-Blochet.

By the mid-nineteenth century, Duvault-Blochet had ascended to a position of political influence and become a titan of Burgundy's wine industry. He was elected to the Conseil Général de la Côte d'Or, the department's governing body, and was regarded as the most influential proprietor of the best vineyards in the region. All 329 of his acres of vines were considered among the top-growth vineyards in the region.

His power and his vineyard holdings were not things he had aristocratically waltzed into. On the contrary, what Duvault-Blochet, a barrel-chested man of integrity and grit, owned he had acquired by being, as one family member wrote, "a nervy old man" unafraid of risk and, for that matter, death.

Family legend has it that Duvault-Blochet once made a business trip to London in the midst of a cholera epidemic. Within a matter of days, he contracted the disease. The manager of his hotel quarantined him in the cellar and summoned a doctor. After examining the patient, the doctor whispered his prognosis to the manager. Thinking that the Frenchman would not understand him, the physician said, "He's done for—too bad, because he's got a constitution of iron." The doctor did not realize that Duvault-Blochet also had acquired a mastery of *anglais*. Irritated by such underestimation of his fortitude, Duvault-Blochet called for a pail of boiling water, a steel brush, and soap. He set about scrubbing himself until he bled, until he was cured.

During the second quarter of the nineteenth century, wine prices plummeted. Cellars throughout the Côte d'Or were filled to capacity, magnificent wines without buyers, and Duvault-Blochet had made a daring gamble. He had put up for credit all he had to his good name for a loan for 10 million francs and he had bought up just about all of the vineyards in the Côte d'Or. His vineyard empire included considerable patches of jewels like Richebourg, Grands Échézeaux, and Échézeaux. It was very informed speculation. In 1854, after three years of record-low wine production, prices soared. Duvault-Blochet was in a position to make the greatest vineyard acquisition of his or anyone's life.

In November 1869, he purchased the Romanée-Conti vineyard. By then, he was eighty years old and even the iron man had come to accept he was mortal. Although he knew he would not have much time left to enjoy the prestige of Romanée-Conti, he had not been able to resist the opportunity to own it, and passed it along with the rest of his estate to his children and their children. Among them were Marie-Dominique. She did not, however, inherit her great-grandfather's genetics.

In 1909, Edmond and Marie-Dominique had their first son, Henri, Aubert's father. He was a healthy baby, a rather plump berry, in fact, and the birth was without any complications. A year later, Marie-Dominique was again with child. She and her unborn contracted diphtheria. The child, Jean, was born and survived thanks to Edmond's mother, who cared for the baby while Edmond did what he could for Marie-Dominique.

Whenever Edmond wasn't caring for his gravely ill family, he was tending to the family's vineyards. When he married into Duvault-Blochet's legacy, Edmond did not have much in the way of viticulture experience. He did, however, recognize the value of the vineyards Duvault-Blochet had assembled.

In the wake of Duvault-Blochet's death in 1874, his holdings had been divvied among the family, some of whom were inclined to sell off the vineyards to the highest bidder—family or not. Edmond could not abide the fractionalization of the family's holdings. He formed a partnership with his brother-in-law, Jacques Chambon. Together they strategized; they appeased and bought out cousins in order to keep the vineyards united in the family, equally divided between the two of them, under one domaine—a domaine that Edmond named after its crown jewel, Domaine de la Romanée-Conti (DRC).

Because Jacques had no desire to manage the day-to-day running of the Domaine, in 1911 Edmond became something of a father for the third time. He took on the role of *gérant*, director, of the family's vineyards. Edmond did not view the responsibility as a job; it was more of a sacred duty.

To help him manage the DRC, he hired a retired army officer. On paper, Monsieur Clin was an odd choice to be sure. Clin didn't know much more than Edmond about tending vines. Edmond, however, went on instinct. He liked that Clin had been an organized, disciplined, and committed French army officer. Clin was a man who had done everything in his power to protect the men on his watch. Edmond astutely sensed that Clin would apply the same dedication to the vines. What's more, Edmond had rightly gotten the impression that in the unlikely event Monsieur Clin would ever slack, Madame Clin would smooth out his wrinkles.

In the summer of 1914, with baby Jean still weak and Marie-Dominique even weaker, the Domaine's grapes were ripening into what promised to be a spectacular harvest. Balancing all this, Edmond undoubtedly realized, would test him in ways he didn't imagine. Then something more unexpected occurred and Edmond had to leave it all behind.

The Austrian archduke Franz Ferdinand and his wife had

been fatally shot on a stone bridge in Sarajevo. The assassination, triggered by long-festering tensions, galvanized alliances in what everyone thought would be "the war to end all wars," and Edmond went off to defend his country against the Central Powers of Germany and Austria-Hungary. About a year into his deployment, Edmond received notice that his wife had died, at the age of thirty-two.

Edmond had come to accept grieving over fallen soldiers. That was war. But to return home, a thirty-three-year-old widowed father of two motherless sons—that was a hell he had never fathomed. Yet he carried on, very much in a spirit that would have made Monsieur Duvault-Blochet proud, fully dedicated to raising his two sons and safeguarding the Domaine his wife had left him. He couldn't save her, but he could ensure that the Domaine was protected.

"And how are the *enfants*?" Edmond asked Madame Clin.

Coming along just fine, Monsieur de Villaine, she said. You will see. She spoke like a proud parent. She tussled Aubert's hair.

Naturally, there was some talk about Edmond and Aubert's trip from Moulins. Because the Domaine then rarely earned a profit and in fact was a money pit, it was necessary for the de Villaines to continue to work their considerable cattle farm in Moulins, for income to subsidize the winery.

Edmond had a routine when he came to the Domaine: First, he visited the stables. As far as little Aubert was concerned, the horses were the best part of his visits to the Domaine. They reminded him of the American West and the cowboys that so fascinated him. He had thought of cowboys, too, every time he had seen U.S. soldiers riding out of France after rescuing his country from the Germans.

That spring marked two years since the end of World War II, and the Germans' surrender to Allied forces among the war-ravaged champagne vineyards in Reims, the largest city on France's Western Front. Little Aubert admired the spirit and grit of the heroic Americans.

In the Domaine's stable, the horses that tugged equipment and carts to and from the vineyards were massive, muscular beasts. Yet gentle. Aubert saw the tenderness in their eyes when they bowed their heads toward him for an apple slice, when they nuzzled their snouts against his cheek. Petting them, the boy smelled that aroma of horse and earth.

His grandfather's favorite was the one called Coquette. Aptly named, she would seem to bat her eyelashes, always flirting for the last piece of fruit, which his grandfather would always nod for little Aubert to go ahead and feed to her. Edmond would shrug as if to say, *What Coquette wants Coquette gets.*

The horse lapped up the fruit, lips and tongue tickling the boy's palm. Aubert giggled and looked up to Edmond, who delicately placed his hand on the back of his grandson's neck—his way of communicating to Aubert that it was time to leave the horses, and of showing his grandson how much he loved sharing this place with him.

Next they would go to the Domaine's small office, heated by a coal furnace, and Edmond and Monsieur Clin would discuss bookkeeping and other pressing matters of the moment. Then it was to the cellar to taste the most recent vintage from the barrel.

Edmond encouraged Aubert to take a taste. The boy took a sip and thought it was so terrible that he dumped his glass onto the ground. Edmond gave him a gentle slap on the wrist and told him to never throw out a drop of wine; it was too precious.

Then Edmond would walk to the vineyards to inspect the vines. Normally Edmond's inspection of the vines was perfunctory. But extraordinarily unfortunate events had lately transpired and his grandfather had a special interest in checking on the *enfant* vines.

As Edmond and Aubert walked west on the dirt road from the centuries-old stone hamlet into the vineyards behind the Domaine, Aubert could smell the fragrance of honey sweetness that came with flowering. Behind them, the bells in the pointed steeple of the Église St.-Martin chimed, each gong softly lingering into the next, until there were no more. It was as if Burgundy, after an interminably long winter, one that had seemingly stretched over several seasons, years of war, was at long last awakening to a spring of new beginnings. Now things would be able to grow. No more attacks. No more dying. At least, this was Edmond's hope.

On either side of the dirt road they walked was the first vineyard they came upon, Romanée-St.-Vivant. They kept walking. Straight ahead, over the brim of the slightly rising road, Aubert could see the top of the white stone cross, and beyond that the hillside of the *côte*, which the holy men in pointed hoods so long ago had dubbed the "Slope of Gold." It, too, was covered in green. In some of the vineyards the rows were stitched in parallel to the hillside; in other vineyards the rows zippered straight up and down the face of the hill. A patchwork of oddly shaped *climats* of uniform misdirection.

And so it was everywhere Aubert looked.

Sometimes, when he was in the car with his grandfather driving along the back roads that wind through the vineyards,

Aubert would stare out the window; if he stared at the vine rows long enough and allowed himself to become transfixed, he would find that the vines seemed to run alongside the car, as if they had uprooted themselves and were chasing them, like they did not want to be left behind, forgotten.

Little Aubert could imagine the vines calling after his *grand-père*, saying. "Monsieur de Villaine, take me, take me!" But the vines were never left behind, never forgotten. They always went with Edmond. *Les vignes* were always present with his family, coiled in the conversations of the adults. A perpetual source of concern and subject of great affection.

Aubert could not understand what was so special about grape plants. He did his best to ignore those conversations. If such talk became too intense and impossible to ignore, as it often did, he would get up and go elsewhere.

Edmond was lean and wiry, with knobby joints and a physique that bent forward, as if he were about to charge off into something. That was indeed how Edmond preferred to go at the world: straight into it, decisively. Politically speaking, he was a royalist. Not that he advocated for a return to anything like the days of the *Tale of Two Cities*, prerevolutionary *Ancien Regime*. It was more that he thought France benefited from a strong, dignified, and just patriarch.

He liked that General Charles de Gaulle had been moving more formally into politics. Only weeks earlier, riding the momentum of World War II victory and his role as the interim prime minister of the French Provisional Government, De Gaulle had galvanized a new political party, the *Rassemblement du Peuple Français*. The RFP's overarching belief was that France should view itself as a world power and conduct itself accordingly, that France should hold itself to the highest expectations and standards. De Gaulle presented himself as an alternative to the typical politicians. "Deliberation," De Gaulle had said, "is the work of many men. Action, of one man alone."

Edmond applied a similar philosophy to his family's winery, home to the crown jewel of French winemaking. He believed it was his responsibility to be just such a leader of action for the Domaine. He was the monsieur with the plan. A vision for how the Domaine ought to be. Truth be told, Edmond had strong opinions on how just about everything ought to be. When Edmond made his opinions known, his family and the extended family of employees took them as orders and knew not to question "*La Pere.*" One of the few times, if not perhaps the only time, Edmond was challenged was by his eldest son, Henri.

Many years after Marie-Dominique had died, Henri had fallen in love with a very distant cousin, Hélène Zinoviev, whose family had fled Russia in 1918 and settled in England, but visited France often. The Zinovievs were among the last of the boyars. Growing up a member of the Russian aristocracy that served the czar, Hélène was raised like a princess. She was a classically trained ballerina and pianist, and Henri immediately fell for her, and she for him. But Edmond would not allow it.

Complicating matters was the fact that Edmond had fallen in love with Hélène's sister, Olga, and already married her. Edmond thought it would be scandalous if his son married his sister-in-law, and so he forbade it. Edmond relented only after his son was called off to fight in World War II and begged his father to allow them at last to be together before he went off to fight the Germans.

As Edmond walked through the vineyard with his grandson,

he was walking toward his final years. There is a Nietzsche-like truism that is something of a cliché in Burgundy: The more a vine struggles, the stronger it becomes, the sweeter the fruit, the better the wine. If anyone could empathize with the vines of Burgundy, certainly it was Edmond. He might not have come to the Domaine a vigneron, but he had become one. Like the vinestocks planted there in that rocky soil ostensibly never meant for vines, Edmond, too, had endured and managed to flourish in inhospitable environments, between rocks and hard places, through harsh seasons. And when he was pressed, what Edmond gave of himself was elegant and pure and strong, like the finest Burgundy *grand crus*.

His hairline had receded, making his most notable facial features all the more pronounced. His dark, arched eyebrows, high above his eyes, almost connected at the top of his delicate nose. On someone else's face the eyebrows might have seemed sinister. Not so with Edmond. At his most severe, he appeared pensive and concerned. More often than not, his face was creased by an expansive grin, which pushed his cheeks into points. In his eyes there was a vitality, a lightness. Despite all the tragedy he had witnessed, in his eyes there was always the hope that came with the next vintage, even when this was not truly his mood. Just as De Gaulle had done in the face of trying circumstances, Edmond presented an indefatigable and dignified front.

From the ankles up Edmond dressed like many men of his era, which is to say, a gentleman. That morning he wore, as he often did, a white dress shirt, a dark-colored three-piece suit, and a thin, dark tie. He could have passed for a banker or a lawyer from Beaune or Dijon. Then there were his shoes: well-worn, dirt-dusted leather boots, cinched tight. The boots were the only outward clue that Edmond was a vigneron.

In the broadest definition of the term, *vigneron* means

"winemaker," anyone connected with the production of wine. The literal translation of the word is "vine grower." In the true Burgundian sense, however, a vigneron is neither of these things. In Burgundy, vignerons do not make wine. On the contrary, they marry grapevines to soil; they work in a communion with nature to raise *enfants* in the spiritually infused ecosystem of *terroir*, hoping to produce thousands upon thousands of wildly diverse, complex wines. What makes the pursuit of such diversity especially interesting is that although there are more than thirteen hundred varietals of grape, Burgundian vignerons work almost exclusively with only two—the Pinot Noir, and the Chardonnay for white wines.

Technically no French word even exists for "winemaker," because the French philosophy is that man does not make wine, God does. Vignerons merely tend, harvest, press, and vinify what He has provided. They kneel and pray, work and wait.

While the concepts of *vigneron* and *terroir* exist elsewhere in France, no community of vignerons takes all of this more seriously than the subculture, or perhaps, superculture of Burgundian vignerons. These philosopher-farmer-shamans strive to bottle the divine as the divine deserves, convinced that the blood of Christ flows from these veins of the earth. *Terroir* and *vigneron*, in Burgundy, are terms of a religion, and of all the sacraments and rituals Burgundian vignerons hold dear, none is more sacred than the marrying of a vine to earth.

Little Aubert loved to be hoisted into Edmond's lap and listen to his stories. There was the one about the World War I victory parade: Edmond was marching along with the French and Allied armies on the Champs-Elysées. Some friends of his were in the crowd and when they saw him began to shout, "*Vive Villaine!*" Within moments, as Edmond told the story to Little Aubert, other onlookers joined in: "*Vive Villaine! Vive Villaine!*" Edmond continued to march in step, his eyes straight, and smiled.

Although he never said as much to his grandson, for Edmond, watching his fellow soldiers in his formation spot their girlfriends and wives, listening to them call out their love, watching them blow kisses and cry tears of joy before their imminent reuniting and resumed lives together, must have been an extraordinarily bittersweet experience, a reminder that his Marie-Dominique was gone.

Edmond had played no especially heroic role in World War I. He made that clear to his grandson. Edmond had simply been a man among men. The real heroes were the fallen, whose names would be etched in stone on the memorials erected in nearly every French village. But to hear his name, to be recognized as a small part of that campaign, well, that had made Edmond proud. Which made his grandson proud.

The resemblance between the two of them was striking. The mannerisms. The physique. Aubert, too, was beginning to lean. Into what, exactly, was something to be determined. Most obviously they were alike in their faces. Aubert had those same arching eyebrows budding on his face; faint as they were, they gave the impression that the boy possessed a wisdom, or maybe it was a skepticism, beyond his years.

The dominant method of populating the vineyards of France at the time was a viticultural technique called *provignage*: An aging vine that historically had been reliably robust would be pulled down, buried under the soil, such that its shoots would push through the earth and grow into new vines. There could be no doubt that Aubert was a shoot of Edmond's stock.

That was true especially since Aubert's father had been absent for the boy's first six years. By the time Aubert had been born, Henri was already off at war. Through Aubert's early years, his father was someone he had heard stories about, as a soldier, then as a prisoner of war, a mythical character to whom his mother wrote letters. During that time, Little Aubert and his grandfather developed an intense bond. They now could read one another.

That morning as they walked together into the vineyards, Aubert could tell that there was a briskness in his *grand-père*'s step. There was an uncharacteristic anxiousness in his grandfather's manner. Aubert had observed enough over the years to know that there had been much happening at the Domaine to make his grandfather anxious.

Although little Aubert did his best to tune out the adult conversations about the vines, it was impossible to remain oblivious to Edmond's troubles with the Domaine. Little Aubert didn't understand all of what had been going on. What he picked up were facts and events without complete context—and as far as he was concerned, without any connection whatsoever to him.

When World War I began in the summer of 1914, Burgundy had been preparing for harvest. As it mobilized for war, the French government did its best to accommodate wineries, allowing vignerons to delay reporting for duty until after the grapes were picked. Wineries were even allowed to keep their horses until the work was complete. The government needed horses like Coquette to pull cannons and ammunition instead of plows and carts filled with picked fruit. Many of the men who went off to defend France came home injured; there was a shortage of vineyard labor and the vines suffered. That had made Edmond and Monsieur Clin's work at the Domaine all the more challenging.

After that first war the Americans got the idea that drinking alcohol was bad for them and fell into the Great Depression. Wine did not sell. What did sell did not bring much profit. Barrels were stacked in cellars until the next vintage, and then the existing barrels were dumped to make room for more wonderful worthless wine. Wine and money were poured down drains.

The year Aubert was born, 1939, World War II started. Aubert's papa, Henri, went off to fight. The Germans captured his unit and put them in a prison. For a time, Germans occupied his grandfather's home, a farm in Moulins.

During that second war, his grandfather's partner in the Domaine, Jacques Chambon, decided he had had enough of losing money on Romanée-Conti and wanted to sell his half of the Domaine.

Edmond was afraid that like so many other families, they would have to sell off chunks of their vineyards, and for next to nothing. For a time it appeared that the prominent Burgundian family the Drouhins might buy out Jacques Chambon. This pleased Edmond. Maurice Drouhin was the Domaine's largest distributor. Aubert's grandfather and Drouhin got along well. They believed in the supremacy of Burgundy and in sparing no cost to take good care of the vines and produce excellent wines.

Drouhin had long dreamed of owning a piece of Romanée-Conti. However, like Aubert's father, Drouhin was also in a German POW camp. When his wife wrote to inform him there was an opportunity to buy half of the Domaine, Drouhin ultimately told her that with everything made uncertain by the war, they could not risk going into debt for the sake of the purchase.

In 1942 a meeting took place at the Domaine. Most of Aubert's aunts and uncles attended. Because Henri could not attend, Aubert's mother traveled through checkpoints of war-ravaged France to attend and represent his interest. Her views were her husband's views. After the big meeting at the Domaine, Aubert heard, the Domaine became something called a *société civile,* a corporation, and a man who had become rich in the wine trade, the patriarch of another prominent Burgundian wine family, a Monsieur Henri Leroy, was his grandfather's new partner.

Aubert was about five years old when he learned one day that World War II was over. A man appeared in his home and Aubert asked his mother if he was going to stay for dinner. The man was frail and sickly-looking; he appeared in need of a meal. Aubert's mother informed her son that, yes, the man would be staying for dinner, and it was her hope that he would stay for a very long time, because this man was Aubert's father, Henri. He had been liberated.

"Say hello to your father," his mother said, nudging little Aubert toward the man. "Give him a hug."

Aubert did as he was told.

Then came the bugs. *Phylloxera vastatrix*, they were called. They were a mystery because no one knew where they'd come from or how to get rid of them. The bugs were tiny, almost impossible to see with the naked eye. They were everywhere in the vines. Eating the vines. Killing them. They caused pimples on the leaves and chewed the wood. Vignerons had tried all sorts of things to save the vines, to kill the phylloxera, but nothing worked. Some vignerons even flooded vineyards to drown the insects. This seemed to work a bit, but not really.

Vines kept dying.

Eventually the bugs came to Romanée-Conti and the vines began to wither. People told his grandfather he should rip out all of them. People told him he had no choice. Monsieur Leroy agreed. At first Aubert's grandfather refused. The vines were more than 350 years old, he said. They were history, he said. If Jacques-Marie Duvault-Blochet could resist cholera, his vines could endure bugs.

Following much discussion, in 1945, when Aubert was six years old, Edmond, the man Aubert was coming to know as his father, and Monsieur Clin gathered with some hired hands in Romanée-Conti. Coquette was harnessed, and the men and gentle beast set about tearing out the vines of Romanée-Conti.

Freshly removed, the bug-infested vines were piled in a twisted frenzy, the long, stringy roots sticking out every which way, as if the *enfants* were trying to cling to something, to someone as they were dragged into a burn pile on the edge of the vineyard.

Little Aubert heard that while the work was being done, as the *enfants* snapped and cracked in the fire, as the smoke from the burning vinestocks curled to the sky, people came from the village of Vosne, people came from all over Burgundy, and watched and wept. Little Aubert was told that Edmond wept.

Aubert and his grandfather came to end of the dirt road, where it formed a T with another dirt road that squiggled north and south through the vines. The only sign at the crossroads was the cross.

To the north were the Domaine's vines in the top-growth vineyards of Richebourg, Grands Échézeaux, and Échézeaux, and beyond those more top-growth vineyards, and then beyond those, little Aubert could see, the roof of the magnificent castle structure of Clos de Vougeot, the ancient home and winery for the monks who first cultivated the Côte d'Or, and where, for a time, before Duvault-Blochet, the grapes of Romanée-Conti were vinified. The Clos was in need of repair, having been damaged by the bombs from the sky. Beyond the Clos was the village of Chambolle-Musigny.

To the south were the vineyards of La Grande Rue, and two more Domaine holdings: La Tâche and Les Gaudichots, then still more vines off in the distance spilled around the town of Nuits-Sts.-Georges, and the restaurant where Edmond and Aubert would sometimes have lunch. It was called La Croix Blanche, which Edmond chose to believe was named after the white stone cross of the Romanée-Conti vineyard. In Nuits there were also the chocolate shops and bakeries Little Aubert liked.

Immediately behind Romanée-Conti was the even smaller parcel of La Romanée, the smallest parcel in Burgundy. Aubert knew those vines were owned by the Liger-Belairs, the family with the priest who, according to Edmond, had such an excellent palate. It was right about where Romanée-Conti left off and La Romanée began that subtle pitch in the *côte* transitioned into a rather severe degree.

From the road where Aubert and his grandfather now stood to the top of the hill made for a challenging hike of about a quarter of a mile, like walking up the face of a large green wave that would never break unto itself. On the top of the hill were dense woods.

Aubert followed his grandfather to the right of the cross, into the Romanée-Conti vines. The vines were small and thin. While all of Burgundy's vines are considered *enfants*, these were truly babies, newly planted vines. It was hard for the boy to imagine that they would ever grow into the sort of sturdy, adult vines that would produce grapes of any kind, let alone the caliber of grapes he was always hearing so much about. Edmond walked among the vines, stopping to inspect them.

One of the stories Aubert's grandfather had told him was about the monks who were here first. They had put the dirt in their mouths and tasted the differences in the soil; that was how they determined where one vineyard would end and another would begin—the soils had different textures, different flavors.

Edmond knelt and touched a vinestock as gently as he'd placed his hand on the back of his grandson's neck. He examined the leaves. He tenderly pushed away the baby canopies of green and studied the shoots, the buds. Madame Clin was right: These *enfants* were coming along nicely.

Edmond watched his grandson wander the vineyard. Little Aubert was not much taller or, for that matter, much older than this new generation of Romanée-Conti vines. Edmond considered that maybe one day Aubert would be the one running the Domaine. Just as these vineyards had been passed down to his late wife and now to him, just as Edmond would pass them on to Aubert's father, Henri. Perhaps these very vines then would be in the hands of his grandson. Maybe the vines and Aubert would grow and work together to make the world's greatest wine. Perhaps, even, they would do well enough for the Domaine to one day break even.

Fat clouds slid to wherever they were going. Maybe, Aubert was inclined to wonder, the clouds would make it to Paris's big-city excitement. Maybe the clouds would float all the way to the sky above the American West, where the Americans rode into adventure on the open plains.

Aubert and his grandfather began their walk back into Vosne. They returned on the same road that cut between the vines of Romanée-St.-Vivant and would end near the Église St.-Martin. The road was called Rue du Temps Perdue, the Street of Lost Time. To Little Aubert that name for a road into vineyards made a great deal of sense.

CHAPTER 5

A Sick Joke

The only clue as to what occupies the address No. 1 Rue Derrière le Four is the very small—maybe six inches high—*R* and *C* atop the tall red iron gates. Behind the gates is a cobblestone courtyard. Wrapped around the courtyard, a flat, one-story, U-shaped building. Entering the gates, the right wing houses the DRC's bottling operation. Ahead and off to the left a bit is the doorway to the cellar—labyrinthine catacombs that wind beneath and beyond the compound. In the left wing there is the tasting room, sparsely furnished with mission-style chairs and tables, and the business offices. Not far from the red gates is the main entrance.

The door, like a work of art, is set within a wall constructed of flat, jagged stones. It is arched, ancient-looking, vaultlike, and made of thick wood, and understatedly adorned with five horizontal rows of windows. Within each window is a tuliplike design made of iron. It is a front that projects equal parts formidability and tenderness. The sight of it calls to mind images of a monastery, or a kind of precious historical archive, or the meeting place for a secret society, or—because there is just enough of the mystical about it—maybe even a portal in time. In a way, to millions of DRC cult-followers around the world, it is all these things.

In the early evening of Friday, January 8, 2010, the door creaked open and the Grand Monsieur emerged. Stepping into a snow globe of blowing flurries, he tightened his scarf and turned up the lapel of his well-worn corduroy sport jacket, which was fading most noticeably at the elbows. There were distinct traces of little Aubert in the face of the seventy-one-year-old. Those arched eyebrows, once youthful wisps, were now salt-and-pepper colored and as wildly bushy as two caterpillars.

He raised one of those eyebrows in frustration as he flipped through the jangling key chain he had been handed officially, though without ceremony, decades earlier. Monsieur de Villaine was frustrated by some of the changes that come with age. It simply was not so, he thought, that everything gets better with age. In all things there is a pinnacle in maturation and then a dissipation. Even with his finest Pinot Noirs there comes a point when the wine has been kept too long and it begins to fade. It's the same with people. Here he was with a clear vision, informed by decades' worth of wisdom and experience, yet now his eyesight was waning. It was a challenge for him to find the right key.

Voilà! At last he found the key to lock up.

Along with the corduroy jacket and the scarf, he wore scuffed leather boots and a wool Jeff cap cocked just so. The workaday winter wardrobe of the average Burgundian vigneron. The clothes both represented and belied Monsieur de Villaine's status. While he looked every bit the vigneron-farmer he was, nothing about his appearance hinted at his wealth. One of the French gossip magazines he made a point of ignoring had recently listed him among the richest people in France.

With his typical straight-backed dignity and grace, Monsieur de Villaine crossed the courtyard to his station wagon, the lone automobile. Often, his longtime assistant Jean-Charles Cuvelier was the last one to leave for the day. Jean-Charles had recently purchased a beautiful new black Mercedes-Benz sedan, which he parked next to his boss's modest and usually dirty Renault. Jean-Charles and the rest of Monsieur de Villaine's employees had left for the day. The Domaine was now closed for the weekend.

The Grand Monsieur had stayed late. He had called to tell his wife, Pamela, he needed to get caught up on paperwork. Which was true. During the winter months, when the vines were dormant, so was much of the wine industry, and these months made for an excellent time to take care of office work. Also true, as Pamela knew, was that the paperwork probably could have waited, or been done by someone else. Her husband had a hard time pulling himself away from the Domaine. She had been confiding in friends and family that she wished her husband would allow himself to retire.

Her husband was stubborn, but not oblivious. The Grand Monsieur was beginning to prepare not only for when he would be gone from the Domaine, but also for when he would gone from this earth. He had already taken the necessary legal steps to turn over his shares in the Domaine to his nieces and nephews. There would be serious tax issues if he were to die without having tended to this business. *Retirement, death*, he thought, *is there any difference?*

Yes.

And no. As with all things related to the Domaine, for Monsieur de Villaine, passing on his shares was much more than a business decision. In order to ensure that his nieces and nephews understood this very fact, in a matter of weeks he would gather them together at the DRC for a luncheon. He would tell them that these shares represent much more than annual returns; they are pieces of the Domaine and the Domaine is something much more than a market valuation, much bigger than any one person. The shares, he would say, are the fruit born of the sacrifice and labor of many generations that came before them, when there was no money to be made from the wines. The Domaine's current success, he would say, is the result of the work of the holy monks, the dukes of Burgundy, a prince, their very own fathers and grandfathers.

Still, retirement was a topic he tried to avoid discussing. A few weeks earlier, during a dinner Aubert and Pamela shared with dear friends, Jacques and Rosalind Seysses, Rosalind had looked across the table and asked Monsieur de Villaine when he planned to step aside.

Jacques and Rosalind had started at their own esteemed Burgundian winery, Domaine Dujac, in Morey-St.-Denis, a town not far up the road, just when Monsieur de Villaine had taken over the DRC, in the 1970s. The seventies were pivotal and an exciting decade for French wines, and, for that matter, even more so for American wines.

In 1976, a British wine shop owner, Steven Spurrier, orchestrated an unprecedented tasting competition that pitted some of the finest wines of France against the most acclaimed wines of the United States. In results that made headlines and stunned the wine world, a wine from Northern California received the top scores. The Grand Monsieur's own role at the center of that controversial affair was something he never discussed publicly.

Rosalind hadn't asked Monsieur de Villaine the question to be nosy. She wasn't at all trying to pry. For her and Jacques it was natural conversation. The only reason she had asked Monsieur de Villaine about his plans was that Jacques and Rosalind themselves had recently begun to transition out of their winery, entrusting their domaine more and more to their two sons. Monsieur de Villaine knew this. He was the godfather to Jacques and Rosalind's son Jeremy, who thought so much of the Grand Monsieur that he had named his firstborn after him. Jeremy referred to the boy as "Little Aubert."

Upon hearing Rosalind's question, Pamela's ears perked up. She was grateful someone else, someone whom they both considered family, would ask her husband this and get him thinking more in this direction. It was especially fitting that it was Rosalind, as she and Pamela had long shared a unique bond: They were both Americans married to famous vignerons. So far from home, they had found their way in the Burgundy wine world together.

It wasn't that Pamela didn't care about the Domaine or wanted to see her husband give up something that he loved so dearly. On the contrary, she loved the Domaine almost as much as her husband did, not just because he loved it, but because the Domaine family had become her family, too. Pamela had done and would do anything she could to support her husband and the DRC.

The great-granddaughter of Charles Fairbanks, who had served as vice president to Teddy Roosevelt and was the namesake of Fairbanks, Alaska, Pamela had gladly given up her idyllic life in Southern California, for her husband and the Domaine. Truth be told, in French society, where it wasn't exactly unusual for men to have romantic dalliances, Pamela was grateful for such a devoted husband, whose only mistress was a winery.

It was just that Pamela thought it was time—time that her husband focused on his own health rather than the health of the vines. And, yes, as she'd confided to one of her sisters-in-law, now that they were in the twilight of their lives together, Pamela thought it was time for them to be together more, to enjoy their vacation homes in the south of France and Big Sur, California, before there was no more time. Pamela was not surprised by her husband's response to Rosalind's question. Monsieur de Villaine gently waved a hand and said it was not something he wished to discuss. He asked that they please move on to another subject and not return to this one.

Jacques was surprised by his friend's definitive reaction. After all, they were like family. Jacques suspected part of the reason the Grand Monsieur did not want to discuss the topic was that there would be questions about succession at the Domaine de la Romanée-Conti. While Dujac's succession plans were clearly defined and relatively simple, the DRC was mired in myriad family complexities and long-simmering tensions that no one openly discussed, out of respect for Monsieur de Villaine.

When Henri Leroy had purchased Jacques Chambon's half of the Domaine, certain agreements had been made. The de Villaine family continued to maintain the exclusive right to distribute the DRC wines in Europe and the United States, but, as part of the deal, the Leroy family took control of distribution of the wines elsewhere around the world.

Another part of the Leroy–de Villaine partnership outlined a change in the management structure. Each family appointed its own director and these two were co-*gérants* overseeing the Domaine's operations. At an annual shareholders meeting held every December over lunch at the Domaine, the families would gather for a briefing by the codirectors. When necessary on decisions related to major expenditures and management appointments, shareholder votes would be taken. Mostly, the gatherings were nothing but joyous and ran smoothly. However, there were rare occasions when Monsieur de Villaine found himself having to defend his decisions that impacted the family's dividends. He was then about to tear up and replant parcels of vines in La Tâche and Romanée-St.-Vivant. Such a move would impact the revenues, as there would be fewer wines to sell for several years. But Monsieur de Villaine was not about to compromise the long-term reputation of DRC wines for short-term gains.

Although there were these annual votes and other codified bylaws governing the corporation—the *société civile*—the success of the Domaine was ultimately rooted in the family shareholders' ability to defer to one of the two directors as the de facto CEO. For more than three decades that person had been Monsieur de Villaine. In recent months the de Villaines had agreed, or so it seemed, that the Grand Monsieur's much younger relation, Bertrand de Villaine, who had been apprenticing at the DRC, would be Aubert's successor. The matter, however, was not a fait accompli.

Some shareholders wondered if another de Villaine nephew, Pierre de Benoist, was a more natural fit. Further complicating the algebra of it all was Monsieur de Villaine's current and long-time codirector from the Leroy side, Henri-Frédéric Roch: How would he and whomever was chosen coexist? Who would defer to whom? Would Monsieur Roch even stay on? Would there be a complete changing of the guard? (Would Jean-Charles stick around?) If Henri Roch did leave his position, there was a widely circulated theory—or was it a wish?—among the shareholders that his spot would go to Henri Leroy's granddaughter, Perrine. If that happened it might very well drag the Domaine back into the family feud that had threatened to destroy it.

Jacques was privy to the Domaine's circumstances well enough to know all of this—well, almost all of it. Frankly, many in the Côte d'Or had some idea of the extent of the Domaine's familial tensions. For years, those tensions weren't much of an issue. Not as long as Monsieur de Villaine was in charge. But when he stepped aside...

Jacques knew his friend well enough to know that he was a shrewd man who had chessed out all of the possibilities and that each outcome was fraught with so much. Yet at that dinner conversation Jacques sensed just how much it all weighed on his friend.

The Grand Monsieur got into his grimy Renault station wagon. The car's interior was also a bit of a mess. The dashboard, along with everything else, was covered in a layer of fine dirt. Waiting for the engine to warm, he rubbed his hands—slender and sun-spotted. He clicked the remote control attached to the driver's-side visor. At the end of the small courtyard the tall red wrought-iron fences next to the visitors' entrance parted.

As the gates closed behind him, Monsieur de Villaine glimpsed in his rearview mirror to be sure they locked together. Perfunctory routine more than anything else. The Domaine was protected by more than gates. The cellar, with all of its thousands upon thousands of bottles of wine, worth millions of dollars, was fortified with a considerable security system. Monsieur de Villaine never gave a second thought to the idea that his vineyards were unprotected.

The Renault rumbled over the cobblestones in the street, which was only slightly wider than the horse-drawn carriages they had been built so long ago to accommodate. *Rues* so tight they seemed like corridors. One of the earliest records of the Vosne-Romanée dates back to AD 650, known then as Voana, which historians say likely means "forest," after the forest above the slope. Over the centuries, the town was known as Vadona, Vanona, Veona, Voone, and Vone, and finally, the Old French, with the silent *s*, Vosne.

In 1866, in order to market Burgundy as a tourist destination and promote its wines, political and business leaders decided to add the name of each village's best vineyard to that name of the village itself. Chambolle became Chambolle-Musigny; Gevrey, Gevrey-Chambertin; Morey, Morey-St.-Denis; Puligny, Puligny-Montrachet; and so on. The tiny heart of all of Burgundy became Vosne-Romanée.

While the name morphed, the village itself changed little. Depending on how you count the *rues* that weave through the heart of the village, there may be a total of twenty streets of wildly varying lengths, ultimately connected to a town center that consists of the Église St.-Martin, a post office, and a defunct well, which when weather permits is adorned with flowers. Electricity clicked on in Vosne in 1900. Central gas came in 1934. Vosne didn't get running water until 1938. Today the town is home to about 450 people. It's not uncommon for the utilities to go out. Wireless Internet access can be hit-or-miss, mostly miss. The streets appear just as they have since the late nineteenth century, when portions of the town were rebuilt after the Franco-Prussian War; the area required rebuilding again after World War I and World War II.

Each time the repairs mostly strengthened the region, though some nuances would never again be the same. During World War II, the magnificent steel-and-glass roof that Monsieur Eiffel put on the Demeure de Loisy Bed & Breakfast in Nuits-St.-Georges was destroyed by bombing. Though it was reassembled, it was done, Madame de Loisy would sadly explain to guests, with less elegance.

As Monsieur de Villaine made his way home, he navigated the narrow streets densely lined with small houses and great domaines, stone façades chipped and cracked, and topped with shingles—some slate, some shakes. Shutters were closed for the evening. Fireplaces puffed chimney smoke into a brilliantly starry sky. The bells of the Église St.-Martin marked the hour with six gongs.

On the walls of many of the ancient structures he drove by were modern granite plaques, their etched letters painted in with red. When Aubert exited Rue Derrière le Four and turned right there was a marker on the wall to his left. The wall encircled the largest domain in the village. The domaine was behind tall, iron blue gates. The sign read:

Ici Vécut et Mourut
Le Comte Louis Liger-Belair
1772–1835
Lieutenant General des Armées
De Napoleon ier et de la Restauration
Grand Officier de la Legion d'Honneur
Grand Crois de Saint-Louis
Proprietaire du Château de Vosne
Et de la Romanée

[Here lived and died Count Louis Liger-Belair. Lt. General of Napoleon the First's Army during the Restoration. A high ranking officer decorated for his services and owner of the Château of Vosne and La Romanée.]

There had a been a time when the Liger-Belairs owned many of the best vineyards around Vosne, but the family's holdings had dramatically dwindled. A number of Burgundians, especially the residents in the village, had a nickname for the current master of the Domaine du Comte Liger-Belair. Behind his back they called Louis-Michel Liger-Belair "the General."

The General had a reputation for being a haughty classist, as if time-warped from prerevolutionary France. Locals talked about what, according to photos in wine magazines, had become his trademark ensemble of bright red pants, gold-buckled loafers, blue-and-white pinstriped shirt (with the sleeves rolled up two or three turns), and the gold ring with his family's coat of arms.

About a decade earlier, when he was in his late twenties, Louis-Michel had arrived in Vosne, moved into the domaine, which his family rarely used, and taken over direct management of his family's vineyards. During previous decades, his family had leased out the vineyards and the vines and the wines they produced had fallen in stature. The General had reclaimed the vineyards and restored his family name and their wines to the celebrated status they once commanded. He'd been aggressive about promoting his domaine and himself.

At thirty-eight years old, Louis-Michel was already a ranking member of the region's most prestigious wine organization, the Confrérie des Chevaliers du Tastevins, and he headed Burgundy's Pinot Noir Association. He was called "the General" not just because whenever a wine writer would visit his domaine Louis-Michel would inevitably talk about the long line of highly decorated Liger-Belair military officers; Louis-Michel had a reputation for swaggering and barking orders.

A domaine is a winery that bottles its own wines. Louis-Michel's bottle-labeling operation was in a second-story room at the north end of his stately domaine. The window afforded a direct view into the boomerang-shaped alley and the Domaine.

Louis-Michel was often at work in that room and found himself staring at the gates of the DRC with complex feelings.

Louis-Michel recognized that Aubert de Villaine was a *grand monsieur*, but it was no secret that Louis-Michel believed the de Villaines had taken advantage of his family decades earlier. In 1933, when his great grandparents fell on hard times, the Domaine bought La Tâche away from his family for what amounted to the proverbial steal. That Romanée-Conti was the undisputed greatest wine in the world, while Louis-Michel's top wine, La Romanée, which came from the vineyard immediately next to the Romanée-Conti vines, was perceived as a less prestigious wine also greatly antagonized Louis-Michel.

The Grand Monsieur was aware that the General's feelings about him and the DRC were nuanced. Monsieur de Villaine himself had been ambitious in his youth. Young masters of domaines must have ambition, he thought, for ambition drives excellence. The Grand Monsieur also empathized with Louis-Michel's heartache over the history between his family's Domaine and the Liger-Belairs. Monsieur de Villaine suspected he would feel the same way if he had inherited the Liger-Belair legacy.

Truth was, the Grand Monsieur saw a lot of his younger self in Louis-Michel. He respected the young vigneron. It was Monsieur de Villaine who had recommended that Louis-Michel be exalted to head the Pinot Noir Association, which Monsieur de Villaine himself had started. After all, Louis-Michel had restored his domaine and wines to greatness. Louis-Michel had chosen to come back to Vosne; he lived in the domaine. He wasn't some absentee *gérant*. Louis-Michel tended to the *enfants* his family had virtually abandoned. Monsieur de Villaine believed the best of Louis-Michel and saw his ambitions as good for all of Burgundy.

On many nights, possibly this one, as Monsieur de Villaine left the DRC for the evening, the General was often up in his bottling room and watching Monsieur de Villaine drive away from the Domaine that now possessed some of the most renowned vines that had once belonged to the Liger-Belairs, and thereby the storied reputation, that had once belonged to Louis-Michel's family. The General could not help envision a day when he would buy back the vines that the de Villaines had acquired from his ancestors.

The next plaque Aubert passed was on the wall of a striking gray stone prerevolutionary mansion behind a spectacular ornate gate that had oxidized green. Both the building and the garden next to it were wrapped in ivy and in disrepair. It was illuminated by one of the very few streetlamps in the village. In the soft sad light, as snow flurries fell, it appeared like a beautiful illustration from a forgotten fairy tale:

Ancien Rendez-Vous
de Chasse
des Ducs de Bourgogne
et Cuverie
du Prince de Conti

[Ancient hunting lodge of the Dukes of Burgundy and the winery of the Prince of Conti]

This property belonged to the Domaine de la Romanée-Conti, acquired by the *société civile* at Monsieur de Villaine's direction. When asked such things—and he was asked often—Monsieur de Villaine would say that to fully appreciate the wines of Burgundy one must understand the history of the place. The vines and the *terroir* are critical, yes, but so is man. Man and his personality bring as much flavor to the wines as anything else.

The Prince de Conti, his heroics, his revolutionary schemes—why, the legend of how he outsmarted the Madame de Pompadour to win the vineyard alone—was worth the price the Domaine had paid for the historic property. How could the DRC not have seized the chance to own the winery once owned by the very prince from whom the Domaine and its crown jewel vineyard takes its name—the man who gives Romanée-Conti so much of its sensuality and power, its balance and complexity?

Just before leaving the village, the Grand Monsieur passed a white stone mansion roofed in blue tile. It was an understatedly rich beauty that, fittingly enough, could not be ignored. The mansion exuded an aura of power and elegant femininity, as did its name: Les Genevrières, property of Domaine Leroy. Monsieur de Villaine noticed a silver Range Rover parked in the drive. Her assistant—*God help that man*, he thought—was there. Meaning it was very likely she was there.

The Domaine Leroy did not have a plaque. It would not have been unfair to suspect that this was because the domaine's owner, Madame Lalou Bize-Leroy, had refused to install one, thinking that such a marker was too small. Instead, Les Genevrières was painted in large flowery script on the wall out front.

Madame Leroy was one of two daughters of Henri Leroy. Throughout Burgundy, Monsieur Leroy, now long since deceased, was still held in high regard. Aubert de Villaine greatly admired him; he felt that he learned a great deal from Monsieur Leroy and that he owed much to him, including one of the jobs Monsieur de

Villaine had as a young man in the United States. Thus indirectly Monsieur Leroy had played a role in bringing the Grand Monsieur together with his wife.

Leroy's daughter Marcelle Bize-Leroy, who had succeeded him, had created quite a different reputation for herself. She was known throughout Burgundy, really throughout the wine world, famously, or depending on your view, infamously, by the nickname her father had given her, "Lalou."

Lalou sounds like the name for someone who, metaphorically if not literally, skips about delivering cupcakes and smiles. Lalou is the name for a person whom everyone knows well and likes. Perhaps then, at one time, the name suited her. According to her considerable reputation, this Lalou was a rarely seen enigma, mostly respected and absolutely feared. She loomed over Burgundy like Leona Helmsley—the intimidating businesswoman who with her husband owned New York's Helmsley Palace hotel and many other properties and struck terror in her staffers. Helmsley had a reputation for being extremely demanding and not especially generous, rarely if ever giving her staff a holiday bonus; Lalou had a similar reputation.

When the de Villaine side of the partnership handed Monsieur de Villaine the keys to the Domaine and made him their director in 1974, that same year the Leroys appointed Lalou their *gérant*. However, because of things she denied but were proven in court, near the end of 1991 her keys had been yanked away. By order of the court and decree of shareholder vote, by her own sister, Pauline, Lalou was stripped of her management role at the Domaine.

Although she and her family still owned half of the Domaine, Lalou was effectively excommunicated from the DRC. The lingering perception was that she was quite unhappy about how she had been treated. Over the years she had talked openly about how disappointed, meaning offended, she was that she had not been invited to taste the new vintages. Business moves she had made in the late 1980s only lent credence to the speculation that she was obsessed with eclipsing the DRC and establishing her domaine as the dominant winery in Burgundy.

In 1988, she had purchased the domaine just on the opposite side of Vosne from the DRC, Domaine Charles Nöellat. She bought up parcels in the best *climats* around Vosne—Le Richebourg, Romanée-St-Vivant—and she set about producing wines that directly competed with the DRC.

Often in Burgundy, many producers own parcels of vines within the same vineyard. In those situations where Lalou now had vines in the same vineyards as the DRC, according to some of the world's most respected wine critics, certain vintages of Leroy wines were judged to be better than the wines of the DRC. From reading the wine critics, you got the sense that all it would take was a series of underwhelming DRC vintages for Lalou to achieve her wish, and the Domaine Leroy would be universally viewed as the top domaine in Burgundy.

Monsieur de Villaine was aware. He was aware, too, that Lalou was of the mind that her daughter, Perrine, would make an excellent codirector of the DRC. None of this troubled him. Just as pouring wine into a glass and leaving it be, giving it time to breathe, often allows the wine to develop its identity and fulfill its potential, in the Grand Monsieur's time apart from Lalou, left alone to breathe in his own space, he had grown into his role and found peace. He had reached the point long ago where passing the Domaine Leroy no longer made him nauseated. When he and Lalou were at the same wine-related events, as they sometimes were, they greeted one another with smiles and civility.

Running any family business is challenging. Running a family business owned by two families is all the more so. Running the Domaine de la Romanée-Conti and at times appeasing two families had been more than he was sure he could withstand. While the Grand Monsieur now was in good spirits, Lalou had nearly broken him.

In the Renault wagon Aubert made a thirty-minute drive on the highway to a point south of Beaune, the ancient wine capital of Burgundy. He negotiated another half hour's worth of winding back roads through woods and cow pastures, until he came upon the village of Bouzeron.

Bouzeron was even smaller than Vosne-Romanée and without any of Vosne's aura of princely aristocracy. At Bouzeron's main intersection, really the hamlet's only intersection, was the town's ancient and defunct public urinal. Just before that, Monsieur de Villaine turned left toward a stone archway and a black wrought-iron fence. In what looked like handwriting penned in steel wire was written "Domaine A&P de Villaine."

Bouzeron's simplicity was one of the things that had drawn Monsieur Aubert de Villaine to the village five decades earlier. Its simplicity and its distance from Vosne and the Domaine. Monsieur de Villaine believed it was important for him to have a life removed from the Domaine. The drive was sometimes a nuisance, but it provided both a real and a psychological buffer, enabling him to drive into and away from the world of the Domaine. Plus, Bouzeron had splendid if underrated vineyards of its own, wholly unlike those in Vosne-Romanée.

The vineyards in Bouzeron had long been considered to have been the equivalent of fixer-uppers when he arrived. The

Hoboken to Burgundy's Manhattan. That was just fine with Monsieur de Villaine. From the moment he'd seen the landscape, felt it, squeezed the soil between his fingers, the young Grand Monsieur sensed their potential and the freedom they would afford him.

In Bouzeron there were no expectations. He would be free to discover the *terroir* in his own way. There were no pressures other than those he chose to impose on himself. All in all, Monsieur de Villaine had quickly concluded that Bouzeron was the perfect place for him to raise his own *enfant* vines, and for him and Pamela to raise their own *enfants*, the children they so desperately hoped for.

Monsieur de Villaine got out of his car, pushed open the gates by hand, got back in his Renault, and drove through. He parked in a stone barn just inside and to the left of the gates. Two large sheepdogs, Sibelle and Ethan, shaggy like giant mop ends, gathered around him as he walked to the steps of his stone home, where the windows glowed yellow warmth.

Immediately inside the front door, he found Pamela preparing dinner. When they'd first met, she couldn't make toast. Once, back in America, when Pamela had been briefly put in charge of a grill, she nearly set a wooden porch on fire. In Burgundy, she developed into a fine cook. She'd found a group of friends—one in particular, another American expat—who taught her to cook and, as a gift, given Pamela lessons with a chef.

Monsieur de Villaine hung his jacket and hat on a rack just inside the door and approached a stack of mail. Those days he was eager for correspondence regarding the Côte d'Or's candidacy for World Heritage Site status. The program is run by the United Nations Educational, Scientific and Cultural Organization

(UNESCO). Such a designation means a site is one of the most significant treasures in the world. UNESCO had already recognized French landmarks such as Versailles, the Cathedral of Notre-Dame, and the banks of the Seine. For the previous two years, Monsieur de Villaine had chaired a Burgundian committee to gain World Heritage designation for the Côte d'Or. If Bordeaux's Saint-Émilion region qualified, Monsieur de Villaine had no doubt that the Côte d'Or met the UNESCO criteria of being an "exceptional testimony to a cultural tradition or to a civilization which is living"; its vineyards "contained superlative natural phenomena or areas of exceptional natural beauty and aesthetic importance"; and it was one of the "outstanding examples representing major stages of earth's history, including the record of life, significant on-going geological processes in the development of landforms, or significant geomorphic or physiographic features."

For a man considering his legacy, to have Burgundy recognized for all that it was, at least to him and his grandfather and father, as a World Heritage Site, well, the Grand Monsieur might be able to retire feeling that he had fulfilled his destiny, because Burgundy would have fulfilled its own.

There was nothing that evening in the mail from UNESCO, but in the pile he saw a cardboard cylinder, the kind an architect might use for blueprints. It was addressed to him and had been delivered by one of France's overnight delivery services, Colissimo. The ends were capped and thoroughly taped. He opened one end and removed the contents: a small note and large sheet of rolled paper. He spread out the large piece of paper on the table. It was a map drawn on grid paper. Monsieur de Villaine immediately recognized the design. He read the note. It had been typed on a computer and was without a single typographical error:

> You just received a map of a part of Romanée-Conti in which you can see a circle. A few days after the end of the harvest, the medium-size vines which are inside this ring have been drilled a few centimeters under the surface of the ground and a piece of black electric wire has been put into the holes. Why? You will know why in about 10 days. In just enough time for you to realize this is not a joke.

He showed it to Pamela, who felt instantly as if some evil had been opened in their home.

That night, Monsieur de Villaine barely slept. The next morning, he called Jean-Charles.

"Jean-Charles," he said, "this is Aubert."

Jean-Charles thought that was it was endearing and comical that after twenty years of working together, Monsieur de Villaine still felt the need to identify himself.

"I have received a most bizarre package," Monsieur de Villaine said.

Jean-Charles listened to his boss explain the map and the markings and then read the note. Jean-Charles noted that as de Villaine read the words, several times the Grand Monsieur gasped and then fell silent, as if reading the note aloud had made it all the more real.

When the Grand Monsieur was finished reading, Jean-Charles asked him what he thought.

"I think," Monsieur de Villaine said, "it must be some kind of sick joke."

Jean-Charles reassured Monsieur de Villaine that yes, of course, that was probably what it was, even though Jean-Charles was not fully convinced that was the case himself.

Just to be safe, Monsieur de Villaine instructed Jean-Charles to contact their vineyard manager, the cellar master, and his codirector, and tell them to be at the Domaine early Monday morning.

CHAPTER 6

Not a Joke

That Monday morning, January 11, 2010, the senior team of the Domaine de la Romanée-Conti gathered in the winery for a closed-door meeting to discuss the threat delivered to the Grand Monsieur's private residence.

The group was made up of Bernard Noblet, the *chef de cave*; Nicolas Jacob, the *chef de vin*; Jean-Charles Cuvelier, the Domaine's director and Monsieur de Villaine's trusted lieutenant; and the co-*gérant*, Henri-Frédéric Roch. Jean-Charles had provided them with a briefing when he had called each of them and asked them to attend, and so as they gathered around a wooden table the mood was somber.

Henri informed his colleagues that he had checked his mailbox at the Domaine before their meeting and found that he, too, had received what appeared to be a very similar package. The men nodded silently as if they were at once processing this new information. They agreed not to open Henri's package, since undisturbed, if necessary, God forbid, it might be useful to investigators. They turned their attention to Monsieur de Villaine's opened parcel.

The Grand Monsieur spread the map out flat on the table between them and once again, though this time without any discernible emotion, read the note aloud. When he finished, the group stood in silence looking at one another. Although they knew each other very well and could almost read one another's minds when it came to performing their tasks at the Domaine, on this unprecedented and potentially troubling matter, no one was quite sure what the others were thinking. They were, after all, an eclectic bunch; each of them had taken a very different path to the Domaine.

Perhaps even more so than Monsieur de Villaine, Bernard Noblet was practically raised at the DRC. In the midst of World War II, with the Germans seizing much of occupied France's manufacturing workforce to help build their war machines, Bernard's father, André, a typesetter, was spared thanks to Monsieur Clin, who gave the young man work sweeping the winery's floors. In time, Monsieur Clin took the broom boy under his wing as an apprentice *chef de cave*, teaching him everything he knew about all of the operations inside the winery, including the pressing of the grapes and the vinification after the harvest, as well as the barreling and then bottling of the wines. André Noblet produced his first vintage in 1952 and continued to do so for the next thirty years.

Bernard's mother also worked at the DRC. The Domaine's vineyard management policy holds that each of the DRC's seven parcels is divided into sections and a caretaker is assigned to each section to maintain it. While Noblet's father made the wines, his mother had been put in charge of caring for the cultivation of the fruit, maintaining two-thirds of the Romanée-Conti vines.

Growing up, Bernard and his three sisters had helped their mother with the pruning and hedging of Romanée-Conti. In their off hours, the family cared for their own vines, which they owned a few miles to the south, just outside Nuits-St.-Georges.

As a boy, Bernard had decided he didn't like vigneron work. It was always either too hot or too cold or too something. In his Catholic vocational high school, instead of oenology classes he had studied mechanics and after graduation took a job a factory in Nuits-St.-Georges that made parts for machinery that bottled and labeled wine. Funny thing was, the more time Bernard spent inside making parts to package wine, the more he realized he missed being a vigneron filling the bottles. He requested to be hired on at the Domaine and was welcomed back as an apprentice to his father.

Now a gangly, pigeon-toed giant in his late fifties, Noblet was shy but friendly enough, unless you were a pretty girl, and then he was perhaps a bit too friendly. Savvy guests who were fortunate enough to secure an appointment for a tasting at the Domaine knew to bring along a pretty girl because in the event that Noblet was the one doing the pouring, a pretty girl, preferably one in a blouse with a plunging neckline, increased the chances of bigger pours all around.

While Bernard learned at his father's side, the old-fashioned way, Nicolas Jacob represented the best of the modern vignerons. He had graduated from the National School of Agricultural Sciences in Bordeaux with a degree in agricultural engineering, and then from the Institut Jules Guyot in Dijon with a degree in oenology. Making wine was something he had dreamed of doing from the time he was a child.

Like most vignerons of his generation, he went to school to learn the art and the science. And like many of the top students in his program he had applied for a job at the Domaine as a long shot. Nicolas was stunned when Monsieur de Villaine showed interest in him and when the Grand Monsieur showed up to listen in on his thesis presentation on the dangers of overplanting a vineyard. The theme of his thesis was that a vigneron should never compromise quality for quantity.

Monsieur de Villaine offered the young Nicolas a job to apprentice under the vineyard manager, who would soon be retiring. It wasn't just Nicolas's fine academic marks or his regard for quality above all else; there was a gentleness about Nicolas that Monsieur de Villaine believed was necessary to care for the *enfants*.

In less than two years, Nicolas, who was then in his early thirties but looked barely old enough to drive, had ascended to *chef de vin*. He wore his hair in a crew cut and generally emanated a relentlessly by-the-book, clean-cut demeanor such that he appeared to fully understand he was responsible for the most valuable vines on the planet.

Monsieur Roch had come into his position as co-*gérant* by way of controversy and tragic happenstance. When Lalou was barred from the Domaine in 1991, in accordance with the corporation partnership bylaws of the Domaine the Leroys needed to appoint someone to fill her slot.

With a supportive vote of all of the Domaine's stakeholders at the annual shareholders meeting at the end of 1991, the Leroys elevated one of Lalou's nephews—one of Pauline Roch's two sons, Charles. Only months after he assumed the role, Charles was killed in an automobile accident, and Henri Roch took his older brother's place. Henri was now in his late forties with a gray, stringy ponytail and a goatee. His everyday sense of style was tie-dyed shirts, leather necklaces, and Converse Chuck Taylors. Henri was honored to take a groovy ride in the role of very deferential co-*gérant* to Monsieur de Villaine.

Jean-Charles was perhaps the least likely of them all to be standing in that room. Beyond the fact that he was born in the Burgundian village of Curgy and was raised to appreciate a good *vin*, the Cuvelier family had no ties to the culture of vignerons. Jean-Charles's father was a laborer. His mother was a teacher. In his youth, Jean-Charles was a skilled enough rugby player that he had reason to think he might achieve his dreams of playing the sport professionally. His hopes were shattered along with his tibia in a rugby match.

With encouragement from his mother—she thought Jean-Charles was very good at coaching the children in the local rugby league—he became a teacher. For the better part of twenty years he taught in the big city of Dijon, ten miles north of Vosne. At the Collège Le Parc, he worked with at-risk students. Because of his working-class background he felt a kinship with the tough, defiant kids. The especially hard cases that intimidated other teachers didn't intimidate Jean-Charles.

Vigilant for ways to connect with his students, to get them interested in school and show them it mattered in their lives, in 1985 Jean-Charles turned to computers. He had an idea that these new machines and those who knew how to operate them would shape the future. He thought computers might give his kids a shot at earning a living. Never thinking how computers might change his life, Jean-Charles learned everything he could and worked the school system to install computers in his classroom. The computers indeed proved to be an engaging tool. As he gained more and more training to pass along to his students, he became something of a computer programming expert.

It led to interesting side work. Jean Charles's brother-in-law and sister-in-law worked in the wine auction business; they knew Jacques Seysses was looking for someone to help the Domaine

Dujac modernize its office administration. In 1991, Jacques hired Jean-Charles to install computers at Domaine Dujac. While Jean-Charles was there, training the staff how to keep their records of the harvest yields, profit-and-loss statements, and orders on the computers, he mentioned to Jacques that it was his dream to taste the wines of the Domaine. The two men hit it off, and by way of an additional thank-you, Jacques asked his dear friend Aubert if he would welcome Jean-Charles for a tasting.

A few weeks later, Jean-Charles was invited to bring a few friends to the Domaine. He arrived at the red gates of the DRC with a handful of his old rugby teammates. During their playing days, they could be brutal on the field, as evidenced by Jean-Charles's snapped tibia, but off the field his teammates had a very civilized love of fine wine. Jean-Charles could hardly believe the experience Aubert and Bernard Noblet gave them.

They started off in the cellar tasting the 1990 vintage, still in barrel. Bernard then opened a bottle of 1970 Grands Échézeaux, then a bottle of the '64. Then Aubert gave a nod, and Jean-Charles and his pals looked at one another in astonishment as Bernard filled their glasses with a 1975 Romanée-Conti! And then the 1983 Montrachet! Jean-Charles still had the wonderful finish of the Montrachet in his throat a few weeks later, in January 1992, when he answered a call from Monsieur de Villaine, who said he had an unusual request.

The Grand Monsieur wanted to know if Jean-Charles could build him a robot of the Prince de Conti. Every year, in January, Vosne-Romanée hosts a village-wide feast of St. Vincent, the patron saint of wine. Many of the villages hold a similar festival to help raise money for a fund that helps support injured and retired vignerons. Monsieur de Villaine thought it would be entertaining for the children if there were a robotically powered, life-size, talking Prince de Conti.

Jean-Charles had to restrain himself from laughing. He very politely explained to Monsieur de Villaine that his request was as unrealistic as it was thoughtful. Jean-Charles explained that engineering the mechanical man was a bit beyond his abilities—a bit, as in impossible—never mind the fact that he would have only about a week to build the thing. The Grand Monsieur figured as much, but he thought he would ask.

A few weeks later, Jean-Charles received another call from Monsieur de Villaine. After hearing from Jacques how well things were going with the computers at Domaine Dujac, Monsieur de Villaine now wanted to modernize his office system, too. In the early months of 1992, Jean-Charles wired up the Domaine and persuaded the very reluctant secretaries that there was nothing to fear in these twentieth-century devices.

Jean-Charles learned that the Domaine was changing in other ways, too. He'd heard that Monsieur de Villaine's co-*gérant*, Madame Leroy, had recently left, or rather had been made to leave, the Domaine. Although he had a new co-*gérant* in Henri Roch, Monsieur de Villaine's workload was increasing and he needed an assistant, someone reliable, someone who would always be there to help him and the Domaine stay organized. Jean-Charles asked Monsieur de Villaine if perhaps he was right for the job. Monsieur de Villaine thought he was.

Now—gathered with the Grand Monsieur and his colleagues around the strange note and map—Jean-Charles was in his late forties and two decades beyond his rugby-playing prime. He still had the husky, squat physique of a baller. The bifocals dangling from a string around his neck hinted at his professorial past. He was widely recognized as the gatekeeper not only for Monsieur de Villaine, but for the Domaine itself. If you wanted to request a visit to the Domaine; if you wanted to request that Monsieur de

Villaine attend an event; if you wanted to petition the Domaine to donate a bottle of wine for a charity auction; if you wanted to try to get on the exclusive approved list of private buyers or retailers, you went through Jean-Charles. Those who worked closely with Jean-Charles sometimes jokingly addressed him as "Dr. No."

After Monsieur de Villaine finished reading the note, the men around him took turns contemplatively rubbing their chins and leaning over the map. Jean-Charles raised and then lowered his bifocals, and then raised and then lowered them again. They all commented on how meticulously crafted the map was.

Sketched on the graph paper it was almost exactly to scale of the vineyard, and what's more, every single one of Romanée-Conti's twenty thousand vines was represented. One of the men remarked that not even the Domaine itself had a map this detailed. Whoever had done this had spent a great deal of time studying the vineyard. It would seem there was also a high probability that whoever it was had spent a great deal of time *in* the vineyard.

Beyond that, none of them said much of anything. They waited for the Grand Monsieur to comment first. Monsieur de Villaine said that upon initial read the note indeed seemed menacing, especially when combined with the map, but when you considered what was really being said in the note, there was nothing much at all. There was no clear threat and so what was there to be afraid of? Monsieur de Villaine made clear he didn't see any point in alerting the authorities, because there was nothing to alert them about. He reiterated his opinion that it was a cruel hoax. To what end he wasn't sure, nor did he care. As was often the case, his staff deferred to his instinct and agreed.

It was settled then. There was nothing to worry about. They would all get on with their day and get on with the business of 2010. In the unlikely event there really was another letter to come, well, they would deal with it then. And so it was business as usual—for eleven days.

On January 20, 2010, Henri Roch found that a second tubular parcel had arrived at the Domaine. Just like the last one, it was delivered by Colissimo. Once again Roch decided not to open it. Jean-Charles phoned Monsieur de Villaine, who was away for the day. When the Grand Monsieur returned to his home the following day, he found that there was a similar package waiting for him in Bouzeron. The Grand Monsieur sat down and, with hands trembling, opened it.

Once again the tube contained a map of Romanée-Conti meticulously sketched on graph paper, and a note. Again the note had been typed and printed out on a computer. This time, however, the map and the note were much more elaborate—this note even included footnotes.

> For the last year, an operation has been under way in Romanée-Conti. It has been a two-part operation. The first part of the operation was done to demonstrate that the second part of the operation is real.
>
> On the map of the vineyard you see a circle. A few days after the end of the harvest, the medium-size vines which are within that circle were drilled a few centimeters below the surface of the ground so that they can be easily found during *buttage* in the vineyard. Pieces of black electric wire were stuck inside those holes. This was done to demonstrate that the operation described below is credible. Because this first phase was only a demonstration, those vines have been not been drilled to be concealed.

For your information, it took only took one night for a team of six to drill them.

The second part of the operation was done to create patterns (or crop circles) in the vineyard (1). This time, after the leaves fell, the vines were drilled much more carefully. That is to say that they were drilled deeper at the level of the roots. And those holes have been filled with various toxic products (2). Either liquids or powder. Then they have been plugged. Plugged with putty and hidden with paint. This operation was much more precisely executed and the operation lasted six nights. At the beginning of March, when the sap will rise (last year that started around March 1) it will carry and diffuse the product within the vines and the vines will die. And crop patterns will appear at the time of the pruning.

SUMMARY:

The target is the reputation of Romanée-Conti. You could try to solve the problems yourself during *decavaillonage*. In the middle of winter it is not obvious. You could also alert the gendarmerie. All they will be able to do is try to arrest the person who will come to get the money. Even if they succeed, you will not be better off. Because that person will not know the placement of the poisoned vines. If you wish save the reputation of the vineyard you need to neutralize the poisoned vines before the sap starts to rise.

Know that it is possible to solve the problem in a fast, efficient, and discreet manner. There are not many ways to solve the problem. There is a trick to quickly find the holes. If you want to know those ways to solve the problem it will cost you 1 million euros. In comparison to what this vineyard brings you...In the future, there will not be any further or higher demands. At the beginning of February you will be contacted one last time at Rue Derrière le Four to give you a location to deposit the money. Have it ready in a sports bag.

This operation has been able to succeed because the vineyard is not protected. If this operation had been done to exact revenge or because of jealousy, the vines which are within the circle would have simply been cut.

Be careful. You have not received all of the information and some explanations are vague on purpose.

1. See "crop circle" in Google.
2. See the two vines at the top of the vineyard.
3. See page 68 of the Revue du Vin de France December 2009/January 2010.

Monsieur de Villaine read the note and examined the map over and over again, trying to break them down, match them to the footnotes—footnotes! Trying to make sense of the circle and two X's in the upper left-hand corner of the map.

Black wires. Painted plugs. Holes just above the surface. Holes drilled in the vines deeper beneath the surface. He couldn't stop his mind or slow his heart. What sense was there in all of this? What exactly was this person—or was it people—trying to convey? What was the threat? What had been done? What hadn't been done? The more he read it, the more it terrified him. The horror was in what was made specific and what was left vague.

What immediately and clearly struck Monsieur de Villaine was that whoever wrote this knew viticulture. *Buttage.*

Decavaillonage. These are the terms of the vigneron. *Decavaillonage* is the springtime chore of churning up and opening the soil around the base of the vinestocks to maximize the flow of air and water to the roots. It's done by rake, by rototill, by tractor, but frequently in Burgundy, as was the case at the Domaine, by horse-drawn plow.

Buttage is essentially the opposite. A task performed in the fall, it means to "earth up" the soil around the base of the stock in order to protect the vinestock from the winter cold and frost. This person, this evil being or beings—Monsieur de Villaine thought—clearly understood the botany and cycles of the vines to know that it is in early March that the sap begins to rise. What's more, whoever was behind this knew the exact date of last year's sap-rising.

They had accounted for every inch of the Romanée-Conti vineyard. They had noted every single one of the vines in the rectangular vineyard. Judging from the number of marked vines on the circle of the map that they had drilled... what?... Hundreds? Thousands? God knew how many; it was impossible to tell.

Had the vines been poisoned in the circle at the center of the vineyard or had he prepared them to be poisoned? If he had poisoned the vines already, was the poison seeping into the soil and contaminating the *terroir*?

Breathe, Monsieur de Villaine told himself.

Slow down.

Read it again.

Slowly.

Slower.

This note seemed to imply that the circle on the map represented the approximate position of vines marked with black wire, and these had not been poisoned, only marked to show that they could have been if that had been the intent, and to prove the "second part": that indeed these vines had been poisoned and amounted to an incipient "crop circle" of dead vines, but vines that could be saved...

...With what?...An antidote?

Is this even possible?

The note—footnote 2—seemed to indicate that a good place for Monsieur de Villaine to start in assessing the credibility of the attack would be to examine the two marked vines at the southwest corner of the vineyard. One thing Monsieur knew for sure: This was no joke.

Almost immediately after reading the second note, Monsieur de Villaine made three phone calls. The first was to Jean-Charles. He gave Jean-Charles a breathless overview of the contents of the package and unequivocally summed up by saying that the vines in Romanée-Conti might be dead and dying, and that the Domaine was being extorted. Monsieur de Villaine called an old friend in the French Ministry of the Interior, gave him a summary of what had been transpiring over the last couple of weeks, and asked for assistance on how next to proceed, stressing his hope that his call would be handled expeditiously and with extreme discretion. And Monsieur de Villaine called Pierre-Marie Guillaume.

CHAPTER 7

Devastator of Vines

Pierre-Marie Guillaume is a meek, bespectacled man with what might be about as round a face as human genetics will allow. If you didn't know any better, you might have guessed he was the cheerful high school calculus teacher loved by all of his students, or pegged him as the guy who every year wore the ugliest Christmas sweater because he sincerely believed it was festive.

Like his father and grandfather before him, Pierre-Marie lived in Charcenne, about a hundred-mile drive northwest from the Domaine. Charcenne is a sleepy village. Its heyday happened around the fourteenth century when the emperor gifted a castle, the Château de Gy, to the archbishop of Besançon, the same year that the town's population hit its peak of three thousand citizens. Now only about nine hundred people lived in the town, which was about as quiet as the Château de Gy, which is now a museum—open only on Sunday afternoons. Suffice it to say, not too much happened in Charcenne, which was fine by Pierre-Marie.

He didn't typically receive emergency phone calls. In fact,

near as he was able to remember, the call he got from Monsieur Aubert de Villaine on January 25, 2010, was the only such call he had ever received from a client, or, for that matter, from anyone.

Pierre-Marie had been doing consulting work for the Domaine for nearly two decades and he knew Monsieur de Villaine quite well. Every holiday, the Grand Monsieur sent his family two bottles of one of the latest vintages, typically La Tâche. (Pierre-Marie would open one and cellar the other.) Ordinarily, when Monsieur de Villaine would phone Pierre-Marie he would begin the conversation casually and inquire about his wife and kids. Not this time.

"Pierre," Monsieur de Villaine said right away, "we are worried. Something strange is happening." Without pause, as if to head off a question, Monsieur de Villaine continued: "I cannot tell you any more over the phone. Can you please come and see us at the Domaine as soon as possible, maybe tomorrow?"

Pierre-Marie knew Monsieur de Villaine to be a reserved man, unflappable and direct; he had never been one to be easily alarmed. And so Pierre did not ask Monsieur de Villaine any questions, nor did he mention that he already was booked with appointments for the next day. Perhaps a bit eager to step into a mystery, Pierre-Marie pushed his glasses up onto his nose, straightened in the swivel chair in his office, and answered that, of course, he would be there.

Like his father and grandfather before him, Pierre-Marie was the owner-operator of Guillaume Pépinières, a nursery. With a degree in agronomy and another in oenology, Pierre-Marie was not only an expert on grapevines; in the world of viticulture he was akin to J. Craig Venter, the father of the Human Genome Project. He served as a consultant to the finest wineries in France and beyond.

Officially, Guillaume Pépinières went into business in 1895, but really it got its start years earlier, when *Phylloxera vastatrix* invaded France and spread around the world. The Grand Monsieur's worst fear was that Romanée-Conti might be in similar terrifying circumstances.

The *Phylloxera vastatrix* phenomenon began in 1863, but it was not until the summer of 1866 that it was formally recognized, in St.-Martin-de-Crau, what was an otherwise insignificant vineyard in the south of France, along the Rhône River. That year, in midsummer, leaves on a number of the vines abruptly turned from a healthy green to alarming blood red. Within weeks, the originally affected vines became an epicenter from which the discoloring spread to neighboring vines, in what quickly revealed itself to be a rapidly expanding circle of death.

Following discoloration, the vines with the blood-colored leaves withered; the grapes, which until then had been progressing nicely to the final ripening of *véraison*, shriveled and dried. As was discovered upon further inspection, so did the tips of the roots. Within a year, the entire five-hectare vineyard of St.-Martin-de-Crau was dead, and the circle of death spawned more circles of death, and those circles created more circles, and so on.

Astutely realizing what was brewing down in the Rhône, Jules-Émile Planchon swiftly traveled to the region from the University of Montpellier, where he worked as a physician and as a professor of pharmacy and botany. Planchon studied the ground-zero vineyard and vineyards in the surrounding Rhône area. He recorded a vivid description what he had witnessed:

> *Everywhere the gradual invasion presented the same phases: after a latent period, some isolated points of attack appeared; during the course of the year, these local points enlarged themselves.... At the same time, multiplication by new foci—advance colonies thrown to distances of several leagues around the centers developed the preceding years; in a word the radiating aggravation of an already confirmed evil.*

Initially, it seemed that whatever *la nouvelle maladie de la vigne* was, was confined to the six departments, or regions, of the lower Rhône Valley, where the grapes were considered a lower-quality fruit that produced "drinkable" table wines.

But then the bloodred leaves spilled into Bordeaux's Midi region, where it was becoming apparent that this plague was beginning to dramatically affect the economy and upend the lives of families. One vigneron from the Midi wrote this account:

"One downsizes equipment and material, lets people go, reduces expenses. One retreats into oneself as in a depression. The beast wins everywhere. In its wake solitude invades all the land. And the horizon takes on an unfamiliar aspect, made up of empty and desolate space. As a palpable sign of the plague, one sees all along the roads, huge carts overladen with dead vines, leading one to a funeral pyre."

Despite these stark realities, because the dying vines were still mostly confined to southern France, Planchon was unable to galvanize much concern about the occurrence. Nevertheless, he was convinced it was a plague and that it needed to be understood as quickly as possible. He needed to discover the source.

A local agriculture board agreed and supported Planchon and a team of two others to unearth the cause. They began, as logic would dictate, with the dead vines. They examined the shoots, the foliage, and dissected the roots. They came up with nothing. Then they figured, why not look at a neighboring vine that was healthy? Despite the objections of the vigneron of the vineyard, they pulled up a healthy vine and put their magnifying glasses over the roots—and were astonished at what they found.

As the team filed in their report, beneath the glass they found "not one, not ten, but hundreds, thousands" of tiny yellowish louses on the wood sucking the sap. Over the course of three days, every affected vineyard they visited, in St.-Rémy, at Graveson, at Châteauneuf-du-Pape, among others, they found these insects, *pucerons*, which Planchon named *Phylloxera vastatrix*, meaning "devastator of vines."

It was both an awful and an awesome discovery. Vignerons rushed to the scene and passed around the loupe in order to see for themselves. They celebrated. For at last they believed they knew who their enemy was and they could make it perish. "Our vines can be reborn," one of the local vine growers shouted. "Our ruin is no longer certain. At last, we can defend ourselves."

In reality, a whole new struggle had begun.

Based upon what they had observed in the vineyards and then under a microscope, Planchon's team was convinced that the bugs, which resembled winged termites, were absolutely causing the death of the vines, moving and killing in colonies. The rest of France dismissed the findings.

The contrarian majority did not discount that the phylloxera were a significant discovery; it was that they saw the bugs as a consequence, not the cause. The agricultural and economic and governmental powers elsewhere subscribed to the theory that it was the vinestocks themselves that were stricken with an *enfeeblement*, and this sickness left the vines susceptible to the bugs, and thus the bugs were drawn to them. It was an insidious sick-vine versus *puceron* egg debate that would rage for the next seven years.

While there were many reasons for the resistance to accept the opinion of Planchon's group, some of it rooted in the age-old scientific establishment tradition of hubris, the ultimate explanation for why Planchon's theory had such a challenging time gaining traction came down to regionalism and snobbery. In departments like Champagne and Burgundy, where they had not suffered at all, the thinking was that the vines in and around Rhône and Hérault were simply of lesser quality and/or their vignerons were lesser vignerons; and, Bordeaux, well, the Bordelais for once in their lives just had the misfortune of being in the wrong place and the wrong time. *C'est la vie*.

In 1870, four years after the first case of the plague was reported and two years after Planchon's team discovered the phylloxera, Paris finally began to take *la nouvelle maladie* in the vines seriously. Someone in the Ministry of Agriculture woke up to these facts: twenty-three of France's fifty-five wine-growing departments had been severely impacted; 50 percent of the country's wine production had shut down; and 17 percent of the French population's income was directly tied to wine production.

The French government, which was engaged in the Franco-Prussian War and was beginning to feel the pinch of plummeting tax revenues, did two things: First, it offered a prize of twenty thousand francs to anyone who could find a way to combat—and their phrasing here was wonderfully political, as it sided neither with Planchon nor those who opposed him—"the new malady characterized by the Phylloxera." Second, it recommended that every wine-growing region establish an official commission, what amounted to a town watch for vineyards. Still, many departments, including Burgundy, mostly ignored Paris's call. The Côte d'Or didn't bother forming its local committee until 1874.

By 1869, while Planchon and his opposing theorists still debated, the phylloxera epidemic spread everywhere, even to Burgundy. The vine growers grew tired of talk and wanted remedies. Burgundians now were sufficiently anxious enough that in November, a Viticultural Congress was convened in the Burgundian wine capital of Beaune. Louis Faucon, a respected grower from Provence, appeared and reported that he had had great success flooding his vineyard and drowning the little bastards. However, the two most intriguing and controversial proposed solutions came from guest speakers from Bordeaux.

When Leo Laliman walked into the congress filled with vignerons who had left vineyards that were barren or filled with vine carcasses, he brought with him an array of lush vines with shoots weighed down by robust bundles of berries. He told the crowd that while rows of vines in his vineyard had been ravaged by the devastators and were as barren and as dead as many of their vines were, right alongside those rows he had vines that were as alive and as robust as the vines before him. The difference between the vines was that the ones that had succumbed to the phylloxera were native French vines and the thriving vines were of stock he had imported from America.

As the crowd reacted with the inevitable mix of hopeful intrigue and nationalistic indignation, Laliman made clear that, of course, the wine made from his surviving native vines was far superior. Laliman wasn't interested in generating controversy, at least any more controversy than he had.

By then, Planchon's prevailing theory was that the phylloxera had entered France's viticulture ecosystem on imported host

American vines. It was speculated, though never proven, that the American vines Laliman had been planting down south might very well have been the cause of the whole mess.

The only point Laliman was attempting to make that day, he made, which was that the American vines were resistant to the bug.

The other buzz-generating presentation came from Baron Thénard, a well-known chemist, who said he had discovered an effective insecticide, carbon disulfide. After several experiments of applying the chemical, which had only recently become available to French industry, Thénard had found that injecting it into the ground around the vines, by using a large syringe-like device called a "pal," wiped out the bug and spared the vine.

These three presentations made clear once and for all, at least to anyone with common sense, that Planchon had been correct all along: The cause of the malady was the bugs. The presentations also set the stage for another debate, which morphed into Burgundian class warfare.

In the wake of the congress, the French government made subsidized carbon disulfide available only to syndicates of growers who would pool together their money to purchase and apply collectively in an organized manner. For families like the wealthy heirs of Duvault-Blochet, paying for the carbon disulfide treatments that ultimately spared the Romanée-Conti vineyard was not an issue.

For many smaller growers, the prescribed regimen of treatments was too expensive. Instead, they favored planting American vines, which didn't sit well with Burgundy's top producers, who were adamant that French vineyards should be populated only by French vines.

The day he spoke at the congress, Laliman never mentioned grafting, but it was already under way in parts of France. And as it became apparent that sulfiding did not guarantee success the way grafting did, the popularity of grafting grew, first quietly and then with more fanfare. The grafting process was painstaking and tedious.

Although grafting was nothing new, and dated back to the time of Pliny, the French had no history with it. Everyone was learning as they went. The challenge was twofold. First, there was finding the right combination of rootstock, which is the part of the vine beneath the soil, and scion, which is the plant aboveground. Then there was landing on a consistently successful technique to marry the scion to the rootstock. The collective ignorance was so profound that amateur grafters actually attempted to graft grapes onto Boston ivy.

Pierre-Marie Guillaume's grandfather, it turned out, had a talent for grafting. He started grafting to reconstitute his own vineyard. When neighboring vignerons saw how well his vines were doing, they began to ask Guillaume if he would mind whipping up some vines for them. In no time, *grand-père* Guillaume was grafting so many vines that he figured he might as well make a business of it, a business that he passed on to his son, and his son passed on to Pierre-Marie. The Guillaume nursery became the third-largest vine-grafting operation in the world.

On January 26, 2010, the morning after he received Monsieur de Villaine's urgent request, Pierre-Marie woke early, shifted his Renault into drive, and traveled the hundred miles to Vosne. Along the way he gazed out at the fields filled with dead, dormant sunflowers. Brown and with their heavy "heads" bowed, they resembled monks in solemn prayer. Most dead plants don't inspire such poetic thoughts. Dead vines, for example, are just dead vines.

At the Domaine, as Pierre-Marie listened to Monsieur de

Villaine's briefing he did so with a poker face; then they headed off to the vineyard. They were joined by Jean-Charles and the young vineyard manager, Nicolas. After a brief examination of the two vines in the southwest corner of Romanée-Conti, Pierre-Marie confirmed what the Domaine's senior team had suspected earlier when they had visited the vines.

"They appear to be dead," Pierre-Marie said.

Monsieur de Villaine grunted. For a long while he said nothing. Then he asked if Pierre-Marie would take one of the two vines back to his nursery-laboratory and see if it could be awakened. He asked Pierre-Marie to learn in any case how they had been attacked.

Nicolas and Jean-Charles helped Pierre-Marie shovel around the two vines and tug them from the earth. One of the men had thought to bring along a couple of plastic shopping bags. One vine would go to Pierre-Marie, the other to the police. Pierre-Marie put his vine inside a plastic bag. As they headed back to Pierre-Marie's car, Monsieur de Villaine trembled.

Indeed it was cold that January day, and Monsieur de Villaine may very well have been shivering. However, something Jean-Charles witnessed later that afternoon was an indication that the Grand Monsieur had been shaken by what had been done to his vines and the crop circles of death he feared might appear in Romanée-Conti.

After Pierre-Marie had left that day, Jean-Charles passed Monsieur de Villaine's office and saw him seated at his desk holding his head in his hands. Jean-Charles saw the Grand Monsieur was weeping.

Jean-Charles was not entirely surprised. Everyone in France knew of the phylloxera and what it had done to the vines. Although Romanée-Conti had been spared during that first phase of the plague at the end of the nineteenth century, when the devastators returned in the first half of the twentieth century, Edmond de Villaine and Henri Leroy's Domaine had not been so fortunate. The ring of death killed their vines. After they were torn out and burned in 1945, until the new vines were planted, the vineyard produced no wine for two years.

Now indeed it seemed quite plausible, maybe even provable, that the ring the hand of evil had sketched on the graph paper might very well be another circle radiating destruction—the promised crop circle of death.

The idea that those very same vines that Little Aubert had visited that day so long ago with Edmond, and which Edmond had gazed upon like a proud father looking through a hospital nursery window, the idea that those vines that had grown up with Monsieur de Villaine; the *enfants* that he had cared for as if they had been his children—the idea that they might very well have been murdered...Jean-Charles could see how that would cause the Grand Monsieur to weep.

CHAPTER 8

Born Again

Twenty-three-year-old Aubert de Villaine was on his hands and knees on the wooden floor of a French army barracks. He could hear the wind that had whipped at his face and hung icicles on his lungs now smacking hard against the creaky building, as if to let the men inside know that winter was still out there, roaring, waiting for them to return.

On either side of Aubert rows of bunks were filled with the men of his unit. Many of them deep into desperately needed sleep. Several of them suffering. At least three of them were gravely ill, including Aubert.

The training that his platoon of officer candidates had done during the previous days of that winter of 1962, in and around Narcy, in central France, had been the most grueling experience of their lives. And despite the warnings their commanding officer had received that the base was under a quarantine for disease, he had gone ahead and marched them right into it. The decision confirmed what many of the men of Aubert's unit had already come to believe: Their commanding officer was a madman.

As Aubert crawled across the floor, out of the room, and into the hallway, he could barely hold up his head. The pain—it felt as if his brain were a knot being pulled tighter and tighter into itself—was excruciating. He was burning up. Covered in sweat. His muscles, even as he moved, felt as if they were turning to wood. He vomited, which led into a coughing fit that drained more of what little energy he had left.

There was a ringing in his ears. His vision was blurry. He was having trouble seeing what was immediately in front of him. Yet he was beginning to see clear visions. Before his eyes appeared the cross and the Romanée-Conti vineyard shrouded in mist. In it he saw a little boy and old man. It was himself as a child, Little Aubert, and Edmond. It was a day like that day with his grandfather. The stables. Coquette. The vines. His grandfather kneeling among the vines of Romanée-Conti while Little Aubert wandered about picking up stones.

From this perspective, Officer Candidate de Villaine now saw something that Little Aubert did not see. Everything was enveloped by an aura, a palpable energy that traveled from sun to vine, from vine to his grandfather, from his grandfather to the boy. In the moment there was—and there was no other way to put it—a divine presence. Nothing and everything was happening.

Officer Candidate de Villaine crawled toward this vision of his past, but once more it was gone, dissolved into his present view of the stretch of corridor before him.

Down at the other end, Aubert saw a softly backlit doorway.

He dragged himself toward it, toward the infirmary.

An aspirin. He had it in his head that an aspirin was what he needed.

Under the lights of the infirmary someone rushed to him, helped him up.

Probably meningitis, a voice said. Another case.

Someone put him into a bed.

Next thing he knew officer candidate de Villaine was seated in a small wooden boat, adrift in his past.

Where am I? he thought

Is this death?

The water appeared to be a pond...

A pond?... Yes. A pond surrounded by woods.

There were familiar smells: the sweet aroma of *pain au chocolat* and roasted almonds. Familiar sounds: He heard the faint background noise and honks of traffic. Much more immediate was the sound of laughter. Above him balloons floated into a sky with a setting sun. His eyes dropped to beneath the balloons, to the children who had released them. He watched them chase after the balloons, running past the vendors selling the baked goods and roasting the nuts.

He looked at his hands. He was holding oars.

He looked straight ahead. There was someone in this boat with him. His vision was a beautiful young woman in her late teens. She was seated opposite him in the subtly bobbing bow. She was watching him, smiling at him. In her lap there was a picnic basket. He recognized this woman. He recognized this moment. This was a time from not so long ago.

From when he was a teenager. A time when Aubert was a teenager, off at Catholic boarding school in Paris. He wasn't an especially devoted student. Rather than focus on the assigned tasks and memorizing the certainties of scientific and religious dogma, he gave himself over to ambiguities of poetry and the expansive spirituality of the philosophers, ancient and contemporary, like Jean-Paul Sartre.

Sartre was a celebrity of that era. Aubert read his idea that man is "condemned to be free." Aubert was of the mind that Sartre must have chosen his words with some degree of wry irony. The way Aubert saw it, a freedom of choice was just that, *liberté*, which for him meant liberty from the vines; the freedom to pursue all the metropolitan culture Paris had to offer.

Aubert could choose to go to the cafés and watch the people and think and daydream and write poetry. He could choose to go to the Louvre, where he could stand before a painting, lose himself in the colors, and choose the possibilities of interp retation. Or he could choose to go on a date with a pretty girl, to a park as magnificent as any of the works of the masters in the Louvre.

Like this day, in a rowboat on the pond in the Bois de Boulogne. It was Paris's largest park, just to the west of "*le 16e*"—the posh sixteenth arrondissement, one of Paris's most elite neighborhoods. The Bois de Boulogne is more than twice the size of New York's Central Park. The vast woods are legendary as a place of romance during the day, and at night a site of sordid passions and illicit vices.

De Villaine's seventeen-year-old self rowed alongside the banks of a quiet patch of the woods. He leaned forward and took the basket from the pretty girl. He took her soft hand in his and helped her step from the wobbly boat. As she stepped out and brushed by him, he smelled the fragrance. It was just then he remembered he had forgotten something very important.

He looked up at her, standing there against the backdrop of the Parisian sunset.

"One minute," he told the puzzled girl. "I will be right back."

He rowed a short distance to an old man selling flowers. He explained to the man that he was with a girl, someone who was very special; they were having a picnic. Aubert asked the man how much for some flowers. He told the old man he did not have much money.

The old man reacted as if he remembered such a time and such a special girl from his own youth. He told Aubert that he was done for the day; his flowers were expiring and would not last until tomorrow. He said that for whatever change Aubert had in his pocket, the old man would sell him all that he had.

The flowers were *pois de senteurs*—sweet peas, or fragrant peas. Pink and delicate orchid-like flowers. Aubert knew these flowers. They were a favorite of his mother. He knew that they were especially lovely in the morning when wet with the dew that would burn away in the sun's heat.

Aubert kept a few of the flowers bundled to hand to the girl. The rest he sprinkled about the boat. It would be a good idea, he thought, if at the end of their picnic he would row his date home on a bed of *pois de senteurs*.

This was 1957, a time that many decades later he would reflect on as a very different time; when the ritual of courting still mattered; when the youth of his generation savored a slowly unfolding romance, rather than lust for the certainty of quick conquest. It was a time, as he would one day put it, when he was in love; when he was always in love; when he was in love with idea of love.

These growing pains of his expanding heart are what consumed him as he transitioned from a boy to a young man in Paris. He did not give a thought to what was happening back in Burgundy. There the Domaine was experiencing its own growing pains, but of a very different sort. Tensions between the old and new, between the Leroys and the De Villaines, were beginning to take root.

When Aubert's father, Henri, returned home after World War II he didn't talk much about it. As far as he was concerned, there simply wasn't much to tell. One minute his unit had been preparing to engage the Germans; the next they were surrounded. Every day of the three years he spent in the stalag was the same: miserable. The only sunshine came with the mail, in the letters from Hélène. If he held those letters to his nose, he could smell her bouquet.

Hélène's letters kept Henri informed about the changes at the Domaine, such as that Jacques Chambon had sold his shares to Monsieur Leroy. Mostly she wrote of Little Aubert growing from the toddler Henri had last seen into handsome boy. She wrote that he was so much like his father, and assured her husband that in his absence Edmond tended to Little Aubert with grandfatherly care.

Hélène wrote almost exclusively of those two things: the *enfants* in the vineyard and their *enfant* son. To write of anything else, she feared, would risk drawing the attention of the Germans, who they knew read every POW's mail. The Germans looked for anything that might be construed as the resistance. The vines and Little Aubert, these topics were pure, and purely their own.

During Henri's time in the prison—that period when Henri and Hélène, who had waited so long before they had Edmond's permission to be together, could not be certain they would ever see one another again—the vines and Little Aubert were the two things that kept them together as one.

Other men in Henri's stalag were not so lucky. One fellow prisoner—and this was a story Henri did share with his family, and he shared it because it offered a rare moment of gallows humor—well, this fellow's wife would send her husband wonderful packages filled with homemade treats. Delicious cookies. Which the man always shared with Henri and other men in the prison.

One day, well beyond their first year in captivity, instead of the package of cookies this man received a letter from his wife. This time she said she was afraid she had some bad news: She was pregnant and wanted to know if her husband could forgive her. The man asked Henri and his friends how they thought he should respond to his wife's question. The men looked at one another, and after carefully considering the circumstances, advised him to tell her that he would talk about this development when he returned, and in the meanwhile to be sure to keep sending the cookies.

There was another story that Henri shared, this one because there was a lesson it. The Germans gave very few rations, mostly in the form of ball-shaped loaves of bread. The bread was given out infrequently, every couple of days. Henri said that many of the younger men, when they received the bread, would ravenously eat it all at once.

Immediately after, these young men would sit full and satisfied, but then in the coming days they would have nothing to eat. It was feast and then famine, physically and psychologically, because their stomachs were tied to their minds. Every time it was the same. It seemed these men had no memory and surrendered to their stomachs.

On the other hand, the older, more disciplined men, when they received the bread, would eat only a very small amount and would ration the rest to themselves in the days when there was no bread handed out. They would never feel full, but they were never without something to eat. Little Aubert understood the moral: On the days that you have bread, be mindful there will be days when you will not have any.

Because the camp had made Henri thin and frail, when he returned home the doctor advised him to avoid excess activity and strain. As he regained his strength, the doctor said, he ought to gingerly recuperate. As far as Henri was concerned, this did not preclude the vigor he put into growing his family. Within six years, he and Hélène had five more children. In addition to Aubert, next came his two brothers, Patrick and Jean, and then his three sisters: Marie-Hélène, Cecile, and Christine.

None of the children wanted to have anything to do with the Domaine, and none of them felt that more passionately than Aubert, who as the oldest male sensed the unspoken wish of his grandfather and father that he join the DRC's operation.

Aubert's grandfather died in 1950, only three years after he and Little Aubert had visited the newly planted vines of Romanée-Conti. Aubert's father, in addition to his work as a commercial banker, took Edmond's place as the de Villaines' representative *gérant* alongside Monsieur Henri Leroy. There were other changes at this time. Monsieur Clin was retiring. After three decades, he was turning over management of the winery and the vineyard operations to André Noblet, who in 1952, two years after Edmond's death, birthed his first vintage of Domaine wines.

During those years, while Little Aubert grew and left for school in Paris, his father and André wanted to maintain the viticulture traditions and philosophy passed down from Monsieur Clin and Edmond. Monsieur Leroy had other ideas. New ideas.

He was interested in modernizing operations, especially when it came to the types of vines that would be used.

The decades of downturn in the international wine market that came with World War I, the challenges of the Great Depression and Prohibition, and then World War II had made it impossible for the Domaine to make a profit. That is why Edmond had kept his farm in Moulins, and why Henri also worked as a commercial banker. They were taking care of the vines, never demanding that the vines take care of them. Their dedication to the Domaine was about many things, but it was not about money.

But those austere times were over. In the 1950s, a more celebratory feeling emerged. Wine flowed. Wine consumption spiked. Even with market fluctuations, the demand would continue to surge for well over a decade. The French market in particular boomed. Around this time, a typical French resident was consuming some forty-three gallons of wine annually.

Among the challenges vignerons faced in keeping the population's glasses filled was that the health of France's vineyards remained in a precarious state. Even with grafting and new chemical treatments, the vineyards that had been repopulated after the phylloxera epidemic were still young, and vulnerable to more traditional viticulture diseases. The vignerons desire to produce enough wine to capitalize on the thirsty market while also maintaining a healthy and robust vineyard led to the great clonal-versus-massal debate.

Massal was the traditional way of selecting the vines that would replace old or sickly vines. During many harvests, a vigneron would study his vines, looking for those that were among the best—meaning producing the best quality of fruit and most resistant to diseases. After several harvests of study, from those vines that were most consistent and vigorous the vigneron would take cuttings, graft them to a stock, and eventually plant them in the vineyard. It would be as if the new vines were of the same family as the parent vines, but there would be diversity in the numbers, and therefore diversity in the wines, as the grapes mixed in the *cuverie* would be similar but far from uniform.

In the 1950s, cloning was new and gained popularity. Cloning was zeroing in on a very few vines of thousands in a vineyard, maybe even only one or two. Cuttings from only those vines would be grafted and planted and grown over and over again. A vineyard was thus ultimately populated by perhaps only a few years' worth of the two direct genetic clones of the one or two chosen mother vines.

Grafting became more and more the work of specialist nursery-laboratories like that of Pierre-Marie Guillaume's family. Cloning was a strategy, in theory, that took some of the guesswork out of farming. If you had a vineyard planted exclusively with the clones of a finite number of mother vines that were proven to be healthy and produce reliably and a high-quality fruit, then you were more likely to have a good harvest. The weather was out of the vignerons' hands, but cloning was something that could be controlled. However, with all the same grapes in the *cuverie,* the vigneron then was left with less complexity in the bottle.

Ultimately, clonal versus massal came down to choosing between a strategy with the primary goal of reliable production of quality wines for profit—in other words, a vineyard strategy rooted as much in the market as it was in the soil—versus a less predictable form of viticulture that left much more up to natural, or divine, selection—a philosophy rooted in the idea that diversity in the vineyard meant more complexity, mystery, and magic in the bottle. Cloning was informed by science and facts and profit and loss; massal hinged on religion and faith, to nature and the vigneron.

Henri, along with André, insisted that massal selection was the best way, the purest way to produce wine, since it let the *terroir* of the vineyard flavor the wine. Therefore, they said, it was the truest Burgundian way to make wine. In contrast, Monsieur Leroy believed that a vine of clones, shaped by the *terroir*, was Burgundy, too, only with maximized chances for a robust yield and rich fruit. Monsieur Leroy thought that to bank the winery's fate on massal selection was a little like gambling in the casinos of the French Riviera.

Henri, with André by his side, and Monsieur Leroy spent many hours in the vineyards arguing. Henri was of the mind that while Monsieur Leroy talked a great deal about the allure of the complexity and mystery of Burgundy's wines, what he really wanted was uniformity and certainty. Here was Henri de Villaine, the banker, saying the vineyard is not the place to look for a speedy return on investment.

But Henri was not one who enjoyed arguing. As with wine, he believed if he let the situation breathe and gave it time, then circumstances would improve. Perhaps because of Monsieur Leroy's seniority and forceful manner, and also out of fear of the diseases that André was finding on the wood in the vineyard, Henri acquiesced.

The Domaine began to populate the vineyard with a vine that had the name of 140-15. It was a decision that would cause vintages of trouble for the Domaine. The 140-15 produced very fat grapes, which is not especially ideal for Pinot Noir, and what's more, as the Domaine would learn, the fruit would take a very long time to ripen. As if that weren't enough, the berries were more susceptible to rot. By the time the 1965 vintage was harvested it was almost entirely riddled with rot. Most of the vineyard had to be regrafted and repopulated.

What would cause the DRC even more strife, however, was the dynamic of power that had been established in that decision. Regardless of his miscalculation with the clonal selection, Monsieur Leroy had very much asserted himself as the lead *gérant* and thereby established a precedent for his heir apparent to do the same.

If this concerned Henri, there was no evidence he let it show. Compared to the stalag, it was nothing. Besides, during the early 1960s, Henri did not have the luxury of dwelling on the methods of viticulture or the Leroy family's position in the Domaine. He was concerned about his eldest son Aubert, who Henri had been informed was in an infirmary at an army base near Narcy, in a coma.

What is this now?

A dark street?

Paris? Yes.

Where am I going?

The eighteen-year old Aubert entered an apartment building.

Ah, voilà! This is when I get arrested.

He was arguing with a doorman. Aubert was telling him he had a poem and he had to give it to a girl. She lived in the building with an old aunt.

The doorman told Aubert to go home. It was too late.

Aubert explained that it was not too late to deliver a poem; it was *never* too late to deliver a poem. What is time when you are in pursuit of love? Perhaps even a bit intoxicated by love.

It was indeed very late, too late, as the doorman insisted. But

Aubert insisted that the doorman did not understand. He had to get this girl in this building this poem right now. Tomorrow was too far away. The doorman warned Aubert that if he did not leave, he would call the police.

"As you wish," Aubert said, and made clear he was going nowhere. Two policemen arrived and dragged Aubert off to the police station.

"Why didn't you simply throw a rock up at her window?" one of the policeman asked Aubert as they escorted him from the building. Aubert agreed that would have been a good idea. But he explained he did not want to risk breaking the glass and making the aunt angry.

"Yet you thought ringing the door buzzer presented less of a risk of waking the aunt?" one of the policemen asked.

Aubert agreed that was another sobering point.

"Will you let us see the poem?" one of them asked.

Aubert was proud of his poem. Maybe if the police read it they would understand his urgency. And so he obliged.

Under a lamp on the Paris street the two cops stood together and silently read the poem. When they finished reading they agreed that it was a very well-written poem. Still, they would not be letting Aubert make this delivery tonight.

They were, however, sufficiently impressed that they informed Aubert they would not call his parents. They would have to keep him at the station for the night. They could not risk him following his heart back to the girl's building at this late an hour. But, they said, he would be released without incident and all would be forgotten in the morning.

Back at the station, they spent the night playing the old French board game of Horses. The game is played on a small piece of wood filled with holes that form a circle, the track. Each player rolls a die and moves their colored peg horse accordingly. Some of the holes can advance the horse many lengths forward. A bad roll can put the horse in a space that drops him several lengths behind.

"*Un jeu de chance*," one of the cops said, laying out the game, "*même que l'amour.*"

A game of chance. Just like love.

And, not that the thought occurred to Aubert that night, but also just like farming grapevines.

Throughout the evening, as they played the game, the cops shared with Aubert some of the foolish things they had done for love. Sometimes these foolish things had turned out not to be so foolish, and turned out to be moves that had advanced their own horses in life forward. Sometimes, as it had gone for Aubert that night, it was not to be and a young foal must trot home.

In the morning, Aubert stepped from the station out onto the Paris sidewalk and blinked, back into his present, the infirmary.

He turned to his bedside and saw the shape of a man in a white jacket. The man identified himself as his doctor. The doctor explained to Aubert that he had been in a coma for several days. He had contracted spinal meningitis. That was the cause of his symptoms, which were so severe some of the staff in the infirmary didn't think Aubert would survive.

Aubert would forever remember what the doctor had said to him when he awoke: "I sensed a fighter in you," the doctor had told him. "I was not going to leave your side. I was not going to let you die. I had faith."

The doctor informed Aubert that his case was so serious he had notified Aubert's parents and they were coming to see him.

Over the next several weeks, Aubert began to recover. He regained his vision in full. His hearing and his memory would never again be what they had been, but Aubert remembered plenty. He remembered the important things.

Aubert viewed the awakening that came after his near death as a "rebirth." A gift not to be squandered. He was determined to live his life to the fullest, with passion and, most of all, with purpose. And he had a new idea of what that meant.

Aubert was now of the mind that while life was indeed a game of chance, life's course is like that pegboard track: regardless of what die you cast and the places you land, ultimately the starting line and the finish line are one and the same—destiny.

CHAPTER 9

The Obvious Question(s)

There he was on his way to being at the center of it all—seated on a chair in a nondescript room facing two cops, nothing but a table between them. It was just like Jean-Charles had seen in the movies. Only this was real.

It was the afternoon of January 26, 2010, the same day Pierre-Marie had declared that the two vines in Romanée-Conti were dead. When Jean-Charles had walked past Monsieur de Villaine's office and witnessed him, head in his hands, quietly sobbing, Jean-Charles had been on his way to the Police Nationale station in Dijon to give a formal statement. There were a few things he needed to get off his chest.

Once he determined it was necessary to contact law enforcement, Monsieur de Villaine did not call the gendarmerie, the local guys responsible for policing the small towns and handling small crimes. The Grand Monsieur thought the gravity of the situation merited more serious attention.

Plus, the local cops were local, villagers. Everybody in the villages knows one another and they talk. Monsieur de Villaine wanted to minimize the risk of leaks and gossip. If people got to talking, it could jeopardize the investigation. Even more unnerving for Monsieur de Villaine, village gossip could jeopardize the Domaine. Imagine what would happen if people started spreading the rumor—or was it the truth?—that the vines and the *terroir* of Romanée-Conti had been poisoned.

Mon dieu.

Instead, Monsieur de Villaine had called an old acquaintance, Hervé Niel, who he knew had ties to the Police Nationale, France's version of the FBI. For a while, Niel had been among the political elite in Dijon, which is where they'd met, but since then Niel had graduated to Paris, where he was a senior official in the French Ministry of the Interior, which oversaw the Police Nationale.

Upon hearing what had transpired, Niel grasped how serious it was. After hanging up the phone, he immediately took the news to the director of the Police Nationale, Christian Lothion. All it took for Lothion to understand the high stakes and high-profile nature of this one was to hear "Romanée-Conti," "poison," and "extortion" together. He directed Dijon to put its best people on the case, and if the best people weren't available, to make them available.

From that point, the police moved as if keeping time with brisk violin music.

The next day, within twenty-four hours of Monsieur de Villaine receiving the second letter, the commander of the Dijon branch of Police Nationale, Régis Millet, and another officer, Inspector Emmanuel Pageault, were at the Domaine collecting the packages and speaking with Monsieur de Villaine, Jean-Charles, and other members of the management team. Meanwhile, one of the Police Nationale's ransom and hostage specialists, François Xavier, traveled from Paris to Burgundy, where he would remain for the next several weeks.

Two detectives assigned to the case were Dijon-based Inspectors: Pageault, who accompanied Commander Millet on the first visit to the Domaine, and Laetitia Prignot. Both Pageault and Prignot were from the organized crime division and surveillance specialists. Their senior officers made clear to both of them that this case was *très, très importante* and was being watched closely by Director Lothion himself.

Inspector Pageault—"Manu"—was a twenty-two-year veteran. He'd spent his first ten years on the force in Paris, the next twelve in Dijon. He was the son of a cop. During his father's forty-year career, his dad had risen from inspector to a commander like Millet.

Inspector Pageault adored his father. As a kid, Manu watched his father going off in plain clothes, a suit and tie, to fight crime and catch bad guys. Manu would jump at every chance to go with his dad to the district. While his father filled out paperwork, Manu would walk about the office, talking to the cops and hearing about the job. He liked the sense of adventure and camaraderie he observed. He liked the fact that these good guys bonded together catching bad guys. In 1989, when Manu did his rookie training, he did it at that very same Autun District where his father was the boss. Manu's first arrest was on a sexual assault case, and when he made the bust, his dad, Commander Pageault, was by his side.

Autun District was in the community of Saône-et-Loire, Burgundian wine country. His old man knew wine. When Inspector Pageault told his retired-cop dad that he was working a case at the Domaine de la Romanée-Conti, his father could hardly believe it. He was proud such a case had been entrusted to his son and outraged that such a crime had been committed.

His father went on about the evil of attacking something so vulnerable and beautiful; something that was, as his father had put it, an integral part of their French heritage and culture. Upon hearing the details, Manu's dad opined that whoever was behind it seemed to have planned it all very carefully, and that it was an inside job.

Inspector Pageault's plainclothes police wardrobe often included a black leather jacket. At forty-three years old, he was fit and handsome. Although he had a boyish face, a friendly smile, and a gentle way about him, when he flipped his cop switch he was all business and went with his gut. Right away Pageault figured the Romanée-Conti case was likely going to be the case of his career.

He hadn't worked much with Laetitia Prignot. She had been on the force only three years. Whereas Manu was the kind of guy who tackled the job and went after bad guys as if he were playing rugby, thirty-year-old Laetitia was more cerebral. Prignot's mother worked for Orange, the French telecommunications company; her father was a helicopter mechanic. Inspector Prignot inherited her dad's attention to detail, his obsession with fitting pieces together precisely into a whole that could take flight. She was meticulous about gathering evidence and presenting it to prosecutors as a case file that would not crash. A quiet, introspective cop, thin with short brown hair, she approached each case like it was a chess match, seeing across the board to the courtroom, anticipating what evidence would be necessary and how it would be used.

Although she was relatively new to the force, Laetitia had quickly distinguished herself as being very good at the job. An expert in electronic surveillance, phone and wiretapping specifically, she had just finished a weeks-long, major assignment in

Corsica, where she'd been working an investigation of France's largest organized crime syndicate, the Brise de Mer.

Though they hadn't partnered much, Manu and Laetitia worked well together. Each viewed the skills of the other as complementing their own. They communicated a lot with their eyes before they said a word aloud or made a move. Commander Millet had proven that his own instincts were very good when he decided to pair them and make them the day-to-day primaries on the case.

Once assigned to the Romanée-Conti investigation, Pageault and Prignot studied the notes and maps. Insomuch as they could, they considered what was likely to come. The two of them swiftly set up their surveillance. They installed cameras in and around the Domaine. One was positioned at the bend in Derrière le Four and was trained on the front entrance; another camera was installed inside the Domaine, focused on the vineyard.

They put taps on all of the Domaine's phones, including the mobiles of Jean-Charles, Henri Roch, and Monsieur de Villaine. They asked for a list of all of the Domaine's employees, current and former. As far as the police were concerned, and as Manu's father suspected, there was a high probability that at least someone connected with the bad guys was on the inside. Anyone and everyone was a suspect.

That said, one of the things that first struck the two detectives was how open and cooperative the otherwise private Domaine was, even Monsieur de Villaine, who they had anticipated, based on his standing and exalted reputation, might be reclusive and aloof. The Grand Monsieur was more than willing to literally hand them the keys to the entire Domaine.

The day they were installing the surveillance cameras, Monsieur de Villaine summoned Manu and Laetitia into the tasting room. Holding a dusty, unmarked bottle in his hands, he asked if they had ever before had any of the Domaine's wines. Considering a single bottle of the Domaine's least expensive wine was at least half a week's salary, the answer was no. The two cops looked at each rather puzzled and politely said they had not. Laetitia didn't drink much wine at all, and when she did it was mostly only "drinkable" whites. She figured it was best if she now kept that information to herself.

Monsieur de Villaine asked a secretary to grab three glasses and he poured them all some of the red from the bottle. The Grand Monsieur watched them drink and raise their eyebrows and smile. *Bon! Très bon!* They agreed. Monsieur de Villaine smiled. Manu asked what it was. Monsieur de Villaine explained it was 1945 La Tâche. What he didn't say, and what Manu recognized, was that the wine and vintage was one of the most coveted in the world. Manu said he hoped the monsieur didn't open it just for them.

The Grand Monsieur confessed that he had not. He had opened it for a friend who had just visited. Manu asked if it was a special occasion. Sort of, the Grand Monsieur replied.

As the monsieur explained, a friend had visited to tell him that he had been diagnosed with an advanced stage of cancer and that he was dying. Monsieur de Villaine had asked the man what year was his vintage, meaning what year was he born in. It was 1945, and so Monsieur de Villaine went to the cellar and returned with the '45 La Tâche. He opened the bottle, and the two friends sat together, and they shared the wine and their memories.

On January 26, 2010, five days into the investigation, Jean-Charles was in the Dijon office seated across from Inspector Pageault and Commander Millet with some information to offer. Thus far, the elite team of investigators had no leads. Not so much as a fingerprint to go on. Whoever had prepared the maps and notes and mailed the packages, the police had learned, must have done so wearing gloves.

During Millet and Pageault's first visit to the Domaine, they had asked the obvious question, or rather questions: Has the domain ever before been subjected to threats or other specific acts of jealousy or revenge? Can you think of anyone who might have reason or motive to do such a thing?

At the time, all of them had said no.

"Are you sure?" Manu asked. "Nothing?"

"Nothing that I can think of," Jean-Charles had said.

Since then, Jean-Charles had noticed some strange activity. Also, he had given his history at the Domaine serious thought. Upon reflection, he had thought of something—in fact, several things. How much any of these things were relevant, he now said to the investigators, he couldn't know, but he wanted to offer these facts to Manu all the same.

He apologized for not saying what he was about to say sooner, when the police first asked, but he explained that thinking about who might have reason to attack the Romanée-Conti vineyard was like thinking in a whole new way, a way he never dreamed of, because he never conceived that anyone, no matter what, would ever think to do such a thing.

"Yes, yes, that's understandable," Pageault said. "Go on."

Jean-Charles explained that he personally had been a victim of an attack. It had been an attack against his character and integrity. In 1994, shortly after he joined the Domaine, an anonymous call came into the Domaine, received by one of the secretaries. The caller claimed that Jean-Charles had been contacting other domaines and wine dealers, offering to sell the contact information of the DRC's esteemed client list for a price.

At that time, Jean-Charles said, he had filed a complaint with the police because the caller had cast doubt on his professional honesty. Also, Jean-Charles said, because the Domaine's business is very confidential and the Domaine has a very select clientele, he did not want to take any chances.

Manu asked Jean-Charles if he had any idea of who might have done such a thing. Jean-Charles did have an idea. He told Inspector Pageault that he would prefer that this be left out of his report, but since the inspector asked, and considering the circumstances, Jean-Charles would share his hunch that he thought it might have been someone associated with Lalou.

Jean-Charles explained that Lalou had been forced from the Domaine under bitter circumstances and the thought crossed his mind that maybe she was determined to exact some revenge. Jean-Charles emphasized there was no evidence that she or anyone associated with her was responsible for the call; it was just his suspicion. He said there had been only one call, and the police at the time did not come to any conclusions.

"What else?" the Inspector asked.

Jean-Charles also described a moment that struck him as suspicious, and which had occurred just after the police first visited the Domaine on St. Vincent Day, January 22, 2010, at around 1 p.m. That day, when he went with Monsieur de Villaine to look at the two vinestocks that were mentioned in the second package, his group came upon six people at the bottom of the wall of Romanée-Conti. One of the people was taking pictures. When Jean-Charles and Monsieur de Villaine passed by, three of the people rushed into a Renault Clio. Jean-Charles said the car was gray and that he had recorded the number on the license plate. He gave that license plate information to Inspector Pageault.

"Anything else?" the inspector asked.

There was one more thing.

Jean-Charles said that between the first and second mailings, that is to say between January 9 and January 20, 2010—he couldn't remember the exact date—he had had several phone contacts with a Serge Bessanko, who identified himself as a writer and said he wanted to write a novel about a prestigious vineyard. Jean-Charles said this Bessanko asked him about the Domaine de la Romanée-Conti. He asked about its precise dimensions, and then he asked if the vineyards had ever been the victim of bad intentions or acts—considering the fact that it was not protected.

The inspector was listening.

Jean-Charles continued.

On January 18, Jean-Charles said, this Bessanko sent him an email that referred to their phone conversations. Jean-Charles took out his iPhone. He informed Pageault that he was forwarding a copy of the email right then to the inspector, who began to look at it:

Jan 8, 2010

Sir, you asked me by phone to send you my novel. It is not finished. I'm not sure you will like it. Some winemakers give a taste of their wine; I'm going to give you only a taste of my novel. Please find it here; it is based on our phone conversations. To have a few comments from you would be very interesting to me. I usually have friends read my texts. And they don't hesitate to criticize my writing. I always take their criticism into consideration when I edit my writing. In your case, I may not have transcribed properly some words or I could have wrongly transcribed your thoughts. I won't have any problem to correct that. You can write to me or call me. But you can also not answer me. I won't be offended. I have never known what pride means. Let me reassure you since my childhood I have always

> bothered everyone I met. My classmates. My parents. And in the end, with time, very few people stick with me. Just a few true friends. It's a rare thing.
> PS: My text talks about three young Bourguoins on holiday. A young woman that I call "Sol," because she is lonely. A boy called Veilleri, which is to be awake. And the narrator.

Attached to Bessanko's email, in the police file, was a sample of the novel-in-progress. It included dialogue, which grew from his conversation with Jean-Charles:

> No one has ever tried to destroy your vineyard?
> The master of the vineyard seems surprised. No.
> Never? Not even in ancient times?
> No, absolutely not.
> Maybe your vineyard is under surveillance?
> No. Why?
> People are not always very good, she says.
> He agrees. You're quite right. And then he keeps going. Right next to here. There was an abbey. Then he continues after a short pause. People have destroyed it.
> The abbey and not the vineyard?
> Oh, my vineyard is very renowned. People respect what they think is sacred.
> Silence followed.
> You're not from the area? asks the master of the vineyard.
> None of us wants to give him an answer.
> I try to change the subject. What you told us is quite interesting. Then I add this to explain: The reason for my curiosity is I'm writing a diary.
> Do you intend to publish it? he asked me. Seeming interested.

Why would you think that I would not publish?... Would the work of a human mind be less valuable than fermented grape juice.

A little pause and then an answer from the vineyard owner.

It's true. A book you can read it again. Wine you can only drink once.

"Because of everything," Jean-Charles said to the inspector, "I thought it was important to share this with you. Again, I am sorry for not thinking of some of this earlier."

Inspector Pageault said he was very glad that Jean-Charles had remembered these things and had brought this new information to his attention.

CHAPTER 10

Almaden

In the late 1840s, a farmer left his home in Bordeaux and boarded a ship bound for America. Étienne Theé did not set out with grand ambitions to revolutionize American winemaking. Never in his wildest dreams could he have imagined his life's work would galvanize characters and events that would nudge the universe of Romanée-Conti. Theé was an immigrant with gold in his eyes. He traveled to California hoping to strike it rich in the Gold Rush.

In the Santa Clara Valley, which was then about a day's journey south of San Francisco, Theé got a deal on a piece of land on the Guadalupe River, between the town of Los Gatos and the New Almaden quicksilver mine. In 1847 he constructed a home there on a hill with a breathtaking view of the valley and surrounding mountains. Gazing out at the landscape, a visitor of the period remarked that "colors indeed seemed to change like the chromatic scale of hues on tempering steel."

It wasn't long before Theé had experienced enough of the so-called rush to decide that prospecting was a riskier proposition than he could bear, and figured it was more of a sure thing to sell wine to dejected or elated gold miners. After all, whether boom or bust, it's a bullish market for booze. On the hillside beneath his home, Theé planted a vineyard. It wasn't long after that he discovered that conjuring wine from California might be just as challenging as unearthing slivers of gold from it.

It was the Franciscan missionaries of the Catholic Church, in need of sacramental wines, who had brought viticulture to Las Californias in the late 1700s. Following their lead, Theé attempted to cultivate the common "mission" grape varietal, the Criolla, of Spanish provenance. No matter what Theé tried, his wines were characterless. Fine for the consecrated blood of Christ, but not so great for commercial appeal.

In 1852, a stylish Frenchman arrived in the valley and settled near Theé's place. Charles Lefranc was not a vigneron. He was a tailor from Paris's posh sixteenth arrondissement. There's evidence to conclude Lefranc came to California, his trunks filled with needles, thread, and fabric, and his head filled with ideas of selling handmade fashion to the nouveau riche.

Within the valley's tight community of French expats, Theé and the much younger Lefranc became friends. For Lefranc, at least, his fondness for the family might have had something to do with the fact that he and Theé's daughter, Marie Adele, had taken a liking to one another. Upon learning of Theé's viticultural challenges, Lefranc suggested, well, why not import some vines from home?

A shipment of cuttings from Bordeaux was arranged—some Pinot, Sauvignon, Semillon, and Cabernet varietals. Theé and Lefranc grafted the cuttings onto the mission rootstock and repopulated the vineyard. In so doing, the men made history, as these were the first high-quality French vines ever planted in Northern California, and they were among the very first planted in the United States. (The only other documented planting of imported French varietals of the time was much farther south, in Los Angeles. Frenchman, Jean-Louis Vignes, had begun planting whole vines shipped from Bordeaux.)

Of more tangible reward for Theé and Lefranc, their new grafted vines produced very fine wine. Whatever the farmer and tailor had done to stitch those French cuttings onto the mission rootstock and into the California soil, pruning here and trimming there as they matured, had worked. Theé and Lefranc were pleased to discover that these vines produced wines that were—as the French tend to say when they allow themselves to think they might have reason to be pleased with a wine—something quite interesting.

The market agreed. The wines began to sell and a partnership was born. Shortly thereafter, Theé and Lefranc made history a second time when they erected an adobe winery, Northern California's first ever for commercial production. Lefranc named the winery New Almaden, after the nearby mine. They filled the cellar with their liquid gold, stored in oaken casks also imported from France.

The partners joyfully drank from those casks in 1857 to celebrate the marriage of Lefranc and Marie Adele. It was then, with Theé getting on in years, that his son-in-law became the proprietor of Almaden. Lefranc expanded the winery's holdings to seventy-five acres, which produced 100,000 gallons of his varietal wines, which were now winning prizes at fairs and competitions. By 1862, Lefranc was recognized as the preeminent winemaker in the region, representing San Benito County at the first California Wine Convention in San Francisco.

As the wealth of the Gold Rush flowed into San Francisco and the city became more metropolitan and interested in wine, New Almaden garnered a reputation for excellence beyond California. The winery made the papers when President Ulysses S. Grant dropped by for a tasting. Gossip columnists wrote about a raucous party at New Almaden where Anna Held, a famous actress of the time, bathed in the estate's sizable tub filled with champagne. True or not, the tale made for fantastic marketing.

As the winery achieved national notoriety, so, too, did Lefranc. In 1876, he was invited to Philadelphia for the grand celebration of America's first centennial. Dressed in a three-piece suit, impeccably fitted, of course, and holding a cane, the bearded vigneron stood before the Philadelphia crowds and proudly tapped on an oval cask some ten feet high, nine feet wide, and eight feet deep, filled with 3,447 gallons of Almaden's best. On the centennial stage in the birthplace of the nation, Lefranc showcased to the world his Almaden wines, born of the first French vines grown in Northern California soil.

By 1880, Almaden was one of the largest and most recognized wineries in the country, and for that matter, around the world. The estate sprawled over 130 acres, and was now planted with vines not only from Bordeaux, but also from Champagne, the Rhône Valley, Burgundy, and even Riesling and Traminer vines from Germany.

Just when it seemed that the winery was charmed—a true fairy tale of immigrants achieving the American Dream—tragedy struck.

One day, in 1887, the serene work at the idyllic estate turned into fatal chaos. Lefranc, who was by then in his sixties, steered a horse-drawn wagon loaded with his wine, a crate of bottles fell and smashed; the startled team of horses took off possessed. Trying to rein in the animals, Lefranc fell into the whinnying, snorting melee of dust; under the hooves and wagon wheels he was trampled to death.

Lefranc's only son, Henri, who now co-owned the winery with his two sisters, Louise and Marie, took over Almaden, sharing management duties with a twenty-nine-year-old Burgundian, Paul Masson. Masson had been raised in a family of well-established vignerons. He came to the United States for the first time years earlier, in 1878. With Bourgogne in the midst of the phylloxera epidemic, his family's wine business struggling like all the rest, Masson had arrived in California's new wine country looking for work, and possibly solutions to combat the bug.

For two years, Masson hung around the Almaden vineyard, which he had heard about over in France, and observed the unprecedented viticultural success the two Frenchmen had grafted and planted. When he wasn't at Almaden, Masson attended business courses at the College of the Pacific. Just as he had planned, in 1880 he returned home to put into practice what he learned. However, back in Burgundy, Masson found the vignerons were still struggling. He hopped on a boat and returned to Almaden, where Lefranc welcomed him back and hired him on as a bookkeeper.

Following Lefranc's death, Masson married Lefranc's daughter, Louise. They honeymooned in France, and then Masson joined his brother-in-law managing the winery until 1909, when the family endured a second tragedy. Henri was killed in a trolley car accident, leaving Masson the sole director of New Almaden.

Proving early that he was a shrewd businessman, Masson kept the winery alive through Prohibition by securing a special dispensation that allowed the winery to grow grapes for "medicinal" wines. He then led Almaden into a period of great expansion driven by the sparkling wine he began to bottle under the New Almaden label ("sparkling wine" because technically only the Champagne region of France produces champagne). Masson continued to build what would become his remarkable dynasty of sparkling wine and "selling no wine before its time." In 1930 he traded the New Almaden vineyards for a 26,000-acre ranch farther south, near Gilroy, California, where he believed the *terroir* was more suited to champagne grapes.

During the next decade, Almaden lost its way. A bank took over the property, selling it in 1941 to a wealthy San Francisco businessman and remarkably extravagant bon vivant, Louis Benoist. At the time that Benoist purchased Almaden, all that he knew about wine was he liked to drink it. And all that anyone seemed to know of him was that he was the president of something called the Lawrence Warehouse Company and he liked the good life; other than that were the biographical shreds he chose to give.

According to the story of Benoist that Benoist gave a magazine journalist, he was a descendant of French aristocracy. His ancestor, the Chevalier Benoist, who had been the court painter to Louis XIV, emigrated to Canada. As far as Louis Benoist's explanation for his own rise, chalk it up to hustle, smarts, and luck.

Benoist also nonchalantly told the journalist about the time he and his wife, Katharine, were sailing off the west coast of Mexico on their recently acquired ninety-eight-foot ketch, the *Morning Star*, when his crew of paid hands mutinied, forcing him and his wife to anchor off a little town called Puerto Vallarta. Benoist didn't explain the cause of his crew's dispute with the management, but he did say, "That landfall was the luckiest thing that ever happened to us. We found a hotel on the waterfront and liked it so much that I bought it next morning."

For Benoist, exotic excursions and impulse buys were not

unique. His life, again as he himself described it, was spent mostly traveling about the five homes he had in California, occasionally varied by a trip east or to Mexico in his private plane. When Benoist gave this interview in 1959, he had owned Almaden for nearly two decades, during which time he had dramatically expanded the vineyard's holdings, moved a mountain to install a swimming pool, and added a helicopter pad to the estate, used for transporting guests and, naturally, himself and Katharine, between San Francisco and Almaden.

Katharine went to great lengths and cost to have Theé's original house meticulously restored, sparing no cost to maintain the home's original character. Katharine, or Katey as she preferred, would not hear of leveling the slanted floors. Her beloved Louie's contribution to the interior design was to decorate the bedrooms, parlor, study, powder room—really, wherever there was a wall—with paintings and photos and drawings of Napoleon Bonaparte, almost as if he were attempting to assure visitors that he was every bit as French as he claimed.

In 1964, Louis and Katharine welcomed to this Almaden estate a guest who was to stay in their home and work at their winery for a few months. What they knew of the young monsieur was that he was a twenty-four-year-old son of the Domaine de la Romanée-Conti; this was his first trip to America; and he was recently discharged with honor from the French military, where he had nearly died from spinal meningitis.

When Louis and Katharine greeted Aubert at the San Francisco airport, they insisted he please call them Louie and Katey. Although Aubert's English was then still sketchy, he understood. He thanked them for coming to get him and for their willingness to host him. He told "Mrs. and Mrs. Benoist" that it was all so very kind of them. Aubert being Aubert, he decided he would stick with the formal address. Anything less, he felt, would be disrespectful. They tossed his bags in the trunk and were off.

From the moment he laid eyes on California, Aubert was enthralled, always comparing this world to the one he'd left in France. During the two-hour drive south to Almaden, Aubert began taking mental notes for an article he would write for the French wine magazine *La Revue du Vin de France*. He noted they drove on a "classic American highway" that crossed "endless fields of sugar beets and potatoes in exquisite company." As they pulled farther away from the city and deeper into the rugged, mountainous wine country, he felt he was entering "the middle of the open American West."

It was early fall, just before the harvest. Driving onto the grounds of the Almaden estate, Aubert marveled at the green vines juxtaposed against the sunburnt hillside. Once inside the Benoists' home, the wooden screen door having gently slapped closed behind him, Aubert found himself unexpectedly standing on the slanted floors, under the scrutiny of the so many sets of Napoleon eyes; everywhere he looked the General was watching him.

There were several reasons he had come to America. First, convinced that he'd cheated death, Aubert was determined to not squander the second chance he had been given. Visiting the States was on his bucket list. From the time he was a small child and saw the Americans rumble into France and liberate his country, he had always been curious of the place that grew such gritty spirit. America was where the cowboys roamed. It was the home of Teddy Roosevelt and the Rough Riders. Americans had a heart and a passion he greatly admired.

Second, and this was the more official explanation, and the one he gave to folks when he didn't feel like explaining the rest, his doctors had told him that the more outdoor physical activity he did, the better it would be for his recovery. Nothing was more therapeutic than good old-fashioned labor in the outdoors. In California, where the weather was mild, without the harsh winters of France, he could work *en plein air* year-round.

The truth was Aubert had what the philosopher in him would call a new existential purpose. On the other side of his coma, he had begun to sense his life bending back to Burgundy, and Aubert wanted to be sure. He wanted to be absolutely certain it was, in fact, his calling. He figured that time away—his time in the United States—would either alter what he was now coming to accept was his life's course, or would amount to taking the long way home.

Aubert had some idea of who and what was in store for him at Almaden when he arrived. He knew a bit of history—Theé and Lefranc, Masson. Of course, Benoist's reputation preceded him. When he left France, Aubert had known where he was going, but he didn't fully understand that this is where he would be—which was wonderfully fine by him, especially since he was still open to the possibilities of where he might end up.

He watched his hosts flutter about to make their historic nest a home for him: Mrs. Benoist talking about how he must be sure to do this and must see that; Mr. Benoist, with his reddened face, eyes blue and mischievous, flashed Aubert a look and said he had just the girl in mind for him. Aubert could practically hear bossa nova in the sunlight pouring through the windows. He felt as if he had stepped into a contemporary world created by Lewis Carroll or F. Scott Fitzgerald. Aubert in a Wonderland of Jay Gatsby.

Certainly, he was now in the mix of an ensemble cast of fiction-worthy characters who were emerging as some of the most influential figures in wine. Along with Katey and Louie, he was scheduled to visit with Robert Mondavi. He was also looking forward to his time with Professor A. J. Winkler. Much of his American experience, including his time at Almaden, had been made possible thanks in large part to winery stakeholders Frank Schoonmaker and Frederick Wildman, who had helped liberate France from the Nazis and then, one could say, helped liberate Burgundy's winegrowers.

While Theé and Lefranc were the first to plant high-quality French vines in Northern California, Schoonmaker and Wildman were the first to successfully import burgundies for a wide American market. Critical to accomplishing their mission, the two men had undone an oppression put upon Burgundy's vignerons by changing the way Burgundy's wine business had operated for at least two centuries.

Schoonmaker was born in 1905 in Spearfish, South Dakota, a town that was about as far removed from wine, or for that matter, about as far removed from anything, as it gets. His family traveled east, where his father, a classics professor, joined the faculty of Columbia University and his mother emerged as a prominent feminist of the time. Young Schoonmaker seemed destined for a similar urbane path when he was accepted into Princeton University in 1923.

That changed dramatically two years later when he dropped out, telling his dad he wanted to "see the world." Schoonmaker went off and roamed postwar Europe for several years, writing travel guides, like *Through Europe on $2 a Day* and *Come with*

Me Through France, books that would be the inspiration for the travel-guide business Arthur Frommer would start a generation later. Schoonmaker researched France's wine regions on foot and bicycle, and learned to speak French as well as a native. He became equally as conversant in wine. Frequently, between his trips to the rural wine regions he would stay in Paris.

It was that time between the world wars when the city served as a muse and meeting place for countless American expats, writers, and journalists—Fitzgerald, Ernest Hemingway, and Gertrude Stein among them. This was the "lost generation," as Stein had labeled them all, a phrase Hemingway made famous in *The Sun Also Rises*, as if he had snatched a butterfly from the air and pinned it to the page. Being lost, as far as Schoonmaker (and Aubert) was concerned, was a fantastic journey of discovery.

In Le Roy Gourmet, one of the Parisian restaurants that was a favorite among expats, Schoonmaker wandered upon Raymond Baudoin. Very recently, in 1927, Baudoin had started *La Revue du Vin de France*, which in short order became the magazine of French wine. The two men got along well enough that they began to travel France's wine country together.

They both believed America would soon drop their ridiculous Prohibition laws and when the inevitable happened, there would be opportunity to be had. Under the mentorship of the native expert Baudoin, Schoonmaker received a wine education and developed a network of contacts far superior to that of any other American, except, perhaps, Colonel Frederick Wildman.

A generational peer of Schoonmaker, Wildman had an affinity for France that developed in parallel with Schoonmaker's, though in a very different manner. After World War I, Wildman was a young U.S. Army lieutenant stationed in Germany. One of his missions was to travel the wine regions of Europe buying up his selections for the officer's mess hall. He, too, sensed that the repeal of Prohibition in the not-so-distant future could mean big business for the French wine trade.

Fourteen years after the war ended, in 1933, when the prediction came true and Americans had the good sense to terminate its "Noble Experiment" and legalize alcohol again, Wildman and some partners bought the New York–based import firm of Bellows & Company and hired Schoonmaker as a salesman.

Schoonmaker and Wildman started by picking up with an effort Baudoin had begun years earlier, which was to persuade Burgundian vignerons to bottle and sell their own wines. This was a seismic shift from the way the region had operated for at least two centuries.

Since the 1700s, before the French Revolution, the central figure in what was then considered the "modern" wine trade was the *négociant*. These middlemen agents, with either their own capital or staked by financiers, bought the vigneron's wine, racked it, stored it, marketed it, sold it, and delivered it under their own *négociant* label.

The arrangement could be a very good one for the vignerons, especially during lean times. The vignerons received quick payment for their product from a single transaction and were spared the cost and hassle of all the rest of what was required to sell bottles on the market.

However, such a deal wasn't necessarily the best. The *négociant* system left the fate of the vignerons in the hands of these brokers who were far more familiar with the actual market prices. While a *négociant* with integrity was an invaluable partner in profit, brokers didn't always operate in good faith.

Another troubling development was that as the wealth of the *négociants* increased, they could conspire to set prices and perhaps even force a vigneron to sell away their vineyard—a vigneron might be funding himself off his vineyard livelihood.

Even in the best of circumstances, the *négociant* system exacted an emotional tax from vignerons. After laboring so hard in the vines and *cuverie* to birth their wines, the vignerons watched a well-heeled agent walk off with their babies and then do as they wished with them. Almost always, *négociants* blended wines from various vineyards, which diluted the purity and unique quality of a *climat*'s *terroir*, which undermined the very essence of a burgundy, which undermined the very essence of a vigneron. With each sale to the *négociant*, a vigneron might feel as if he were selling away not only his long-term security, but also his soul.

Already, in the early part of the twentieth century, some domaines had begun to bottle their own wines and negotiate their own sales with buyers representing import-export companies and distributors. However, these domaines were very few and tended to be the more established wineries with brand recognition and market demand, such as the domaines of Marquis d'Angerville, Henri Gouges, Armand Rousseau, and Edmond de Villaine. It was under Edmond's man-of-action leadership that the DRC began bottling part of the Romanée-Conti vineyard in the latter part of the 1920s and started selling directly and negotiating their sales.

Schoonmaker and Wildman developed relationships with the big domaines—they got to know Edmond and Monsieur Leroy and then Henri de Villaine quite well. Schoonmaker and Wildman would point to them by way of convincing the rest of the Burgundian vignerons that domaine bottling was the future. The two Americans pledged to the vignerons that if they made the investment in bottling, Bellows & Company could find the market in America.

Of course, the estate-bottling strategy served Schoonmaker and Wildman. The two men were stealing the business away from the *négociants*. The vignerons were fine with that as long as it gave them more control over their own product, provided Schoonmaker and Wildman could deliver on their end of the bargain. That's where everything got a bit sticky, at least for a while.

The market analysis Schoonmaker and Wildman offered was, at best, overly optimistic. During the fourteen years of Prohibition, Americans had grown accustomed to going without—or rather, let's be honest, most of them were happy to drink whatever was available. They weren't exactly savvy or discerning or willing to spend good money when it came to their booze. Grapes were grapes and wine was wine. California, French, Martian, made no difference.

Schoonmaker set out to educate and inspire the American market to recognize the difference and pay for it. He became a one-man publicity machine. In 1933, the very same year he was put in charge of sales for Wildman's Bellows & Company, with Wildman on the ground in Burgundy, Schoonmaker published his first book on wine, *The Complete Wine Book*, which was based on a series of articles he had written for the *New Yorker*. The book became something of the twentieth-century wine bible. That, combined with his magazine work, spawned a wine industry press.

Of all the French wine regions, there was no question that Burgundy was Schoonmaker's favorite. "Heartwarming and joyeux, heady, big of body, magnificent and Rabelaisian, this is Burgundy," he wrote. "The most celebrated poet of Bordeaux, Biarnez, wrote of the châteaux and the wines so dear to his heart in cool and measured Alexandrines reminiscent of Racine.

Burgundy is celebrated in bawdy tavern songs." In other words, if pomp and pretense are your thing, you'll love Bordeaux. If you're looking for passion and one hell of a good time, Burgundy is where the action is. Burgundy also happened to be where the majority of Schoonmaker and Wildman's business was.

Their plan worked. Bellows & Company primed a clientele of a national network of fine wine retailers, elite hotels, the best restaurants, and more, a new market of sales direct to high-end customers throughout the United States. Their business came to a halt with the start of World War II. Bellows was sold to U.S.-based National Distillers; Wildman rejoined the U.S. Army Air Force; and Schoonmaker was recruited into the Office of Strategic Services.

The OSS was the precursor to the Central Intelligence Agency, and it drew, like the CIA would, from Ivy League universities for its recruits. Schoonmaker may not have had an Ivy League degree, but he had a PhD in European wine and his occupation of wine buyer provided the perfect cover: Schoonmaker could travel in and out of occupied territories without raising suspicion. He undertook several missions into Spain and France to aid with the resistance, so deeply covert that the United States ambassador to Spain was often in the dark as to Schoonmaker's whereabouts.

The Spanish police managed to arrest Schoonmaker, but he found a way to slip out of Spain and promptly signed on with the U.S. Seventh Army, which invaded southern France in 1944. Schoonmaker was injured when his jeep hit a land mine and was discharged honorably, and presumably very reluctantly, as a colonel, the same rank as his colleague, Wildman.

With the war won, Schoonmaker and Wildman returned to the United States and to the wine business, experts like never before in the channels necessary to acquire information and wine. Each took his own path back into the business, but those paths intersected at Almaden.

It turned out that all those houses, yachts, planes, helicopter rides, exotic trips, and God knows what else Louie bought had him and the winery running a little low on capital and in need of a serious sales mind to get the place in the black. In the early 1950s, Benoist brought on Schoonmaker as investor-partner, which worked doubly well for Schoonmaker, who wanted a home base from which he could run his own import-export endeavor. With Schoonmaker's guidance, Almaden acquired a 2,200-acre ranch in Paicines, California, planting what would become the largest premium vineyard in the world.

Premium wines just happened to dovetail nicely with Wildman's strategy. After the war, he signed with the company that had bought his Bellows & Company away, National Distillers. Wildman worked for National for a few years before starting his own premium import-wine business, Frederick Wildman & Sons, where, thanks to his long-standing relationships and respect in Burgundy, he landed the Domaine de la Romanée-Conti as a client. Wildman served as the Domaine's exclusive distributor in the United States.

Sure, Wildman said when Henri de Villaine and his partner, Monsieur Henri Leroy, had asked him if Wildman & Sons might have some work in the States for Aubert. The way Henri and Henri had put it to Wildman, Aubert needed some time away and would be coming to America; he wasn't sure if a life in wine was right for him, they told Wildman, but he was curious.

The colonel arranged for Aubert to spend a few weeks working with his offices in New York. Wildman took Aubert under his arm and showed him how the wine business worked inside:

who was who, whom to trust and whom to avoid, the necessary taxes and tariffs, the unnecessary markups. He gave Aubert a master's class in the very network of importing, exporting, and distribution of wines that Wildman and Schoonmaker in effect had created. Not only that, Wildman arranged for Benoist to host and employ Aubert at Almaden, and facilitated meetings for Aubert with a couple of California's most influential figures in wine.

The morning after he arrived at Almaden, Aubert got on with the rest of his American adventure. Benoist had made a guesthouse available to Aubert. It was one of seven Benoist had constructed on the estate. All of them were well-appointed and filled with Napoleons. A handful of the Generals clad in various military ensembles hung around as Aubert dressed the part of a vigneron: a simple plaid shirt, khakis, and boots. He put on his thick-framed black eyeglasses. His dark brown hair was in a military-style buzz cut. He could have passed for an American GI home on leave. He rolled up his shirtsleeves and headed outside to begin a day with a living legend.

Aubert would describe the start of his first day in California wine country in the *La Revue du Vin de France* like the romantic poet he was:

> The door of my room opens onto a big terrace that overlooks the valley and mountains and vines. When I go out, the sunlight hits me like a fist in the face. The bright and shimmering landscape lays in front of me. It is like looking into a painter's palette. Rich in all nuances, pink, red, green, and brown. In the background is the horizon of yellowy-orange, ochre and

blue. The hills which surround the valley are burned by the sun.... At the top of a hill, I see part of a hillside has been simply cut away, to make some space to build a house, a swimming pool, and a garden. I walk onto the grass behind the house. The grass moves toward the vines like the prow of a ship. This morning is like a paradise.

In the very near distance there are bushes of flowers, and just beyond them the first rows of the vines spill from the terrace, and then go straight ahead, cut by only the road at the bottom of the slope. And so it goes with all of the hills surrounding the valley. The contrast between the aridity of these naked hills and the rich deep green of the vines is striking. I take the road that leads down to the buildings where they make the wine and where I will have my first meeting of the day with Professor Winkler. He is waiting for me at the wooden house just next to the road that divides the vineyard into a slope and flat plain.

Professor Albert J. Winkler was the foremost viticulture expert in California, which essentially made him the foremost viticulture expert in the world. A native of Texas, Winkler earned a PhD in horticulture at the University of California, Berkeley, in 1921. He spent the rest of his life working as a specialist in viticulture and winemaking. He was responsible for a number of industry-altering studies and breakthroughs.

In the late 1920s, Winkler developed a sulfur dioxide gassing process that made it possible to ship grapes to the East Coast. In the 1930s, he founded the department of viticulture at the University of California, Davis, where he would be a faculty member for four decades.

One of the first projects he undertook there was to identify which parts of California were best for growing grapes based upon his "heat summation method"—an extensive analysis of the average temperature in an area and its impact on fruit ripening. In what became known as the "Winkler Scale," he determined there were five "zones," I through V, with I being the best (coolest) and the most comparable to Burgundy's Côte d'Or. Only two years before Aubert arrived in California, in 1962 Winkler published the book *General Viticulture*, which immediately became a definitive guide for winegrowers. He served as a consultant for Almaden.

When Aubert approached the professor, he was in his sixties, thin, lean—in fact, very much like a vine himself, Aubert thought. Winkler wore an olive-colored shirt, khakis, and a wide-brimmed olive-colored hat cocked just so, giving Winkler more the look of a Texas cowboy than a viticulturist.

They climbed in an old jeep, similar to the ones that had zipped around the bases of France. Winkler shifted the thing into drive and darted off into the vines. For hours they bounced through the vineyards first planted by Theé and Lefranc, stopping here and there when Aubert noted the pronounced differences between the viticulture of France and California.

Right away, Aubert questioned the distance between the vines. There was as much as six to nine feet between each. In Burgundy it is much less. Burgundians believe that some density of planting is good for the vines because it forces them to compete for nutrients; the more a vine struggles, so goes the cliché in Burgundy, the better the vine and the wine. In the Côte d'Or, a vineyard is bit like an arena of horticultural Darwinism (which is an undercurrent of the massal selection versus clonal debate). Aubert's question was the question of a novice, but that's precisely what he was.

Winkler stopped the jeep and the two men got out for a closer inspection of the vines. While they kicked at the dirt and talked, a filthy flatbed truck rumbled past, bouncing over the dirt, kicking up dust. The bed of the truck was fenced in with walls of wooden slats. Mexican workers were hanging on to the exterior of the slats. Aubert and Winker stopped talking for a moment to let it go by. Aubert was amazed that with all the bouncing none of Mexicans' hats or none of the Mexicans themselves went flying from the truck.

Winkler explained that while France has a Mediterranean climate that is often very wet, California is a desert. California vine growers had tried planting at the same density as Burgundy's vignerons, but due to the lack of water and nutrients, the vines ended up killing each other. Winkler pointed out that despite the Burgundy–Northern California differences in planting density, the volume of production per acre was the same. While there were fewer vines per acre in California, each vine produced more quality fruit.

In Burgundy, while there were more vines per acre, each vine, due to its struggle, produced less useable high-quality fruit.

They stopped at a stretch of parcels "that were like an undulating sea." The previous year, the professor explained, this was all prairie; now it was half Pinot Noir, half Cabernet, and the vines would be harvested in about three years' time. They were planted as cuttings, without being grafted to a stock. Aubert asked about the strategy of planting two different varietals in the same parcel. Such a thing in Burgundy would be sacrilege.

Ever since the medieval monks cleared away the brush, married the Pinot to the ostensibly inhospitable landscape, and birthed excellent wines, no one questioned that planting Pinot and only Pinot in the Côte d'Or was best according to God and nature. Considering vulnerable cuttings were planted, Aubert also asked Winkler about phylloxera.

The answer to the phylloxera question was easy: San Benito County had never been troubled by the pest. Vinegrowers there hadn't grafted in more than ten years. As far as mixing varietals in the same plot, Winkler joked that every French vine grower he'd ever talked to asked the same question.

We have different standards than you do in Burgundy, he said.

Citing his breakdown of the five regions, the professor said that in Northern California varietals ought to be matched to the climate, whereas in Burgundy it was all about *terroir*. Zones I and II, for example, are good for Pinot and Cabernet, while Zone V, where Almaden had vines in Paicines, were best for dessert wines like the golden Muscats. (The fact that the Muscat varietals are so hearty, able to grow in even the most dry regions, explains why they are the oldest grape in the world.)

Aubert was stunned there could be such uniformity of temperature within each region. Winkler assured him the climate was incredibly consistent. In fact, he said, we constantly monitor it.

Winkler walked Aubert to a phone-booth-like shelter; inside he gestured to a thermometer that recorded the temperature.

Winkler explained that his team recorded the temperatures weekly and the numbers confirmed that "all of this valley is Zone I and that we were right to just plant fine grape varietals, like Pinot Noir and Chardonnay. Those grapes ask for a slow maturation, a late harvest. So Zone I, the coldest or rather the less warm, is well suited for them. In Zone V, we harvest in late August. Sometimes we'll harvest in November. Napa Valley, the old wine coast, is between Zones II and III; Cabernet is definitely successful there, Pinot Noir has much more difficulties. This valley of the Paicines is coolest in California; it's hotter in San Francisco, which is on the ocean, due to the altitude."

Aubert could not believe that there was no concept of *terroir* in California.

Winkler reached up and scratched the front of his head, pushing his hat up as he did so, and said, "Here we have around ten valleys that have the climate and soil that allow the culture of fine grapes. None of them are really different from the others except for the climate. As for the soil, within each valley it can't be considered so particular or unique that it gives a specific character to the wine that it produces. Of course, the combination grape-soil-climate does make the wine different. The Pinot Noir from Paicines gives maybe a better wine than other ones on other hills, but, in all the valleys classified I and II regions, the Pinot Noir does well."

Aubert tried one last time.

"You mean," he said, "there will never be any kind of specific, scientific connection between a particular grape and the soil?"

"First," Winkler answered, "we have to find out, get our data, get that experience, the proof that it's really clear that in a certain vineyard, a grape gives a better wine than another grape. The Cabernet, according to the expert-critics, is totally successful in California, whereas the Pinot Noir is still looking for its right land. Maybe we found it here, at Paicines.... Within years, with experience, and research, we will know that maybe the land here will give a better Pinot Noir than the other varietals. If so, we'll set up the vineyard accordingly."

On their way back to the jeep, Aubert was startled when a flock of birds launched from the vines, screeching around him so close he swore he felt wings flutter against him.

"One of the pests we do have to contend with," Winkler said.

Just before he climbed into the jeep, Aubert kicked at the soil. It was covered in splinters of stone that reminded him of Burgundy.

Louis Benoist did take it upon himself to set up Aubert on a blind date with a young lady he thought was compatible with the Frenchman. Benoist even planned the date. A baseball game. Take a pretty American girl to America's pastime, have some peanuts and Cracker Jack, and who knew, Benoist thought, maybe that night Aubert might not come back.

Aubert would never be able to say for sure where exactly the game was played or which teams played. "Play," as far as Aubert was concerned, was a bit of an overstatement. A bunch of guys standing around for two hours, all of them each wearing a big leather glove on one hand, waiting for the guy holding the stick and the guy throwing the ball to make something happen.

Aubert couldn't understand any of it and quickly determined he didn't want to. The girl was a daughter of friends of the Benoists. She didn't seem to know much about this baseball business herself, nor seem to care. She was nice enough, but shy. She didn't say much. The men who held the stick didn't do much, so not much of anything happened. Aubert would remember that night as one of the most boring of his life.

A few evenings later, however, Aubert had the best nights of his time in America. He joined Professor Winkler and a crew of Mexicans on a two-day-long chore. They rode on horseback deep into Almaden's vineyards, where the terrain was especially challenging. Winkler needed to determine if these vines were ready for harvest before the staff went through all the trouble of carting all of the pickers and gear out to these locations.

Aubert couldn't believe it: He was on a horse, with gauchos, riding over the mountains of the American West. Just like an American cowboy. Because they were so far out and had many parcels left to examine in the morning, that night, as planned, they set up a camp. They built a fire and cooked over it. The sun disappeared on the other side of the valley, behind the mountains, then the day pulled behind the ocean.

The moon rose harvest white and strong. Stars appeared. There were many and they were brilliant. One of the Mexicans had brought a guitar. In the cool, clear night, around the fire, they sang. They pulled bottles of wine from their saddlebags and drank until the bottles were empty. They talked about where they had come from, their homes. The Mexicans were fascinated by the idea that Aubert had come all the way from France, and by the fact that his family had such vineyards and such history. The more Aubert talked about Burgundy, the more he realized he missed it.

Another highlight of his trip happened a few days later. It was his appointment with Robert Mondavi. Mondavi was then in his fifties, stuck between all that he had accomplished and all that he was yet determined to do. He was enmeshed in a ruthless feud with his family.

Mondavi's father and mother, Cesare and Rosa, emigrated from Italy in the early part of the twentieth century, settled in Minnesota, and then, in 1923, moved to Northern California. After a solo reconnaissance trip to the region, Cesare, who was in the wholesale fruit business, had decided there was more opportunity in the California grape business. He also had discovered he liked the climate better in California: It was much warmer than Minnesota and he encountered far less prejudice against immigrants.

By the time Cesare's eldest child, Robert, graduated from high school, his family had already gone into the wine business, the cheap, jug-wine sort. Robert (with help from Rosa) convinced his father to seize the chance to buy the oldest winery in Napa Valley, Charles Krug and its vineyards, and get into the business of more upscale wines. The winery was in a state of neglect and disrepair. In relatively short order, the Mondavis made Krug profitable and popular with critics and tourists alike.

Under Robert's influence, the winery focused more on public relations, most noticeable in its unique tasting room. In a grand celebration, when the reconditioned 1914 railroad car was opened at Krug it was christened "Rose of the Vineyard." It was seventy-five feet long, with a semicircular bar at the center and striking views of the Krug vineyards. The Rose is where Robert and Aubert sat and talked. As he shared that day with Aubert, Mondavi had even greater plans. In fact, Mondavi was especially pleased to be talking with Aubert of the great DRC because Mondavi believed California's bright future was in emulating France's history and winemaking traditions, especially those of Burgundy.

Only two years earlier, Mondavi and his wife had taken their first-ever trip to France, visiting the great domaines of Burgundy and châteaux of Bordeaux. The attention that the Burgundians gave to their vines and wines astonished him. Robert saw the attention to quality that was given to the much smaller vineyards and production far exceeded anything he witnessed in Northern California, which, he now came to believe, put too much emphasis on quantity.

One dinner that Robert and his wife had in Lyon—the courses thoughtfully paired with the exquisite wines—confirmed for Robert, as he told Aubert, that California needed to think about wines more as the French did. Robert believed Krug and the Mondavi estate ought to aggressively move in that direction of artisanal, boutique wines.

Silently looming over their conversation that day was also an open secret. Robert was in the midst of a toxic fight with his family, in particular with his younger brother, Peter, and their mother, Rosa. Robert's brother and mother, and his sisters, thought Robert had returned from France with impractical notions and quite full of himself, or rather even more full of himself than ever before. Cesare, who very much had been the dominant patriarch and arbiter of such disputes, was now deceased, and so the Mondavi family's soap operatics raged and the family was unraveling.

Aubert had no idea of just how bitter the fight had become. He didn't know at that very moment that Robert had been effectively banished from Krug and put on a monthly stipend, which would soon be revoked. However, Aubert, like just about everyone else who was in wine or picked up a newspaper, did have a general sense of what was happening in the Mondavi family. Listening to Mondavi talk, Aubert was impressed by his vision and by his energy, but he also saw that Robert wasn't especially nice. He sensed in the man a desire to impose his will, perhaps even at the expense of his family.

Well into his trip to California Aubert picked up the phone to call his father in France. During his time in the United States, Aubert had begun to take stock of what he had learned there, about winemaking and the business of it, but really, he reflected on what he had learned about himself. The United States, New

York and Northern California, was a wonderful place. He loved all of it and he wanted to return, of that he was sure. But compared to France, California wine was like a promising adolescent; the wineries were still trying to figure out what they wanted to become.

Everywhere he turned, everyone he spoke with—Wildman, Benoist, Winkler, Mondavi—looked with envy and admiration to France's rich winemaking history for guidance. Everyone—heck, including some of the Mexicans—told him how fortunate he was to be a part of such traditions at the greatest Domaine in all of France. Listening to Mondavi, Aubert recognized how fortunate he was to have the family he did. The Domaine was a business, yes, but first and foremost it was a family, his family; a family and their ancient *enfants* rooted in much more than a climate zone.

When Aubert's father, Henri, answered the phone, they engaged in the usual small talk and then Aubert said he was ready to come home. He asked his father if he could work at the Domaine.

CHAPTER 11

Leads

On a January night about a week after Monsieur de Villaine received the second package, Inspector Prignot parked her unmarked vehicle on a dark street that was between a row of stately mansions and a tree-lined canal (one of the many waterways of its kind that crisscross Burgundy). Satisfied she had kept a reasonable distance between her vehicle and No. 6 Promenade Aristide Briand, she began to watch and wait.

Prignot was in the village of Genlis, only a few miles from Vosne-Romanée. The mansions on the street were a mix of beautiful residences and run-down converted apartment complexes. No. 6, located behind a long brick wall, most definitely was one of the latter. On the seat next to Prignot was a file, and paper-clipped to the file was a photo of the man she was waiting on, Pierre Leduc. In the picture she'd pulled from his Facebook page, Leduc was in his late thirties, shaved head, with a pencil-thin Van Dyke. She looked at her watch: 7:35 p.m. Based upon their reconnaissance, Leduc was due home shortly. Prignot checked for a visual of her backup: a couple of cops in a car at the other end of the street.

If you had asked Prignot and her partner, Inspector Pageault, how they thought the case was going at that moment, depending on their mood they might have said they'd made quite a bit of progress or not much at all. Thus far, they were well on their way to ruling out that writer Bessanko had any involvement. They'd scoured Bessanko's background, pulling his bank and phone records. There was nothing there. Bizarre coincidence. Almost every case has one, the veteran Pageault reminded his younger partner.

Most notably and troubling, however, the investigators had uncovered that the DRC wasn't the only victim that had been marked. At the Police Nationale office in Dijon, an examination of the two parcels addressed to Monsieur de Villaine had revealed no fingerprints inside or out, other than his own, meaning whoever wrote the notes, sketched the maps, and mailed the packages had done so with gloves, indicating a certain level of sophistication. But the mailing information proved useful.

The tubes had been mailed from a Colissimo branch located in the Gare de L'Est, one of Paris's busiest train stations. Based upon the time stamp on the mailing form on the second of the two parcels, police were able to pull a video from the store's security camera. It showed that the person who had brought in the tube was a man, or at least appeared to be a man, dressed in a dark ski cap and dark jacket. The person was careful to never face the camera—another sign, as far as the inspectors were concerned, that whoever they were hunting wasn't clumsy.

One of the details police could see in the video of the day was that the person had shipped not one, but two tubes. A search of the Colissimo tracking forms showed that the other tube was sent to Chambolle-Musigny, the tiny hamlet immediately north of Vosne-Romanée by two miles. Beyond that, the Paris branch had no more information for the police. For the specific destination address, police would have to check with the post office that served Chambolle and hope someone there remembered. The postal clerk in Chambolle had recalled the package and was relatively certain it went to the Domaine Comte Georges de Vogüé.

Inspector Pageault promptly visited the Domaine de Vogüé, a stone, castlelike building on a narrow street in the center of Chambolle. He found Severine Godine, the domaine's office manager. Godine, who also lived at the domaine, was stunned to receive a visit from an inspector with the Police Nationale. Pageault asked if the domaine lately had received any usual packages. None that she was aware, Severine said. Then again, she informed Pageault the domaine had not yet opened after the holiday and much of the mail was sitting unopened. Pageault said he needed her to look through that mail, now.

Severine found two tubular mailings, just like Pageault had described. One had been sent the same day the mysterious figure was seen on the Colissimo video; the other had been sent a couple of weeks earlier, mailed the same day as the first parcels mailed to Monsieurs de Villaine and Roch. Both of the Vogüé parcels were addressed to the Comtesse Claire de Causans, who co-owned the domaine with her sister, Marie Ladoucette. Severine had seen the packages before but she had not given them a second thought. She'd figured they were one of the many calendars that their vendors send every year.

Sure enough, when Pageault and his team opened those tubes back in the crime lab of the Police Nationale's Dijon branch, they found maps, and notes of threats and the promise of a forthcoming demand, very similar to the ones Monsieurs Roch and de Villaine had received, only here the target was the vineyard called Musigny, which like Romanée-Conti also had the most esteemed and most rare classification of a burgundy—*grand cru*.

Burgundy's contemporary classification of wines was made official in the mid-1930s by a branch of the French Ministry of Agriculture, the Institut National des Appellations d'Origine (INAO), which was created in 1935 for the very purpose of overseeing the hierarchy of French wine.

In theory, the system the INAO codified for burgundies is straightforward. First, they determined which areas within the region merited recognition as a unique appellation. Then within each of those appellations, they determined which specific vineyards commanded an exalted status. The result was a ranking of four tiers. From least to most prestigious, there is *régional*, *village*, *premier cru*, and *grand cru*.

The ostensible simplicity belies a reality of considerable complexities. There's the overarching question of why some areas within a region are recognized with the designation of appellation while others are not. And why is it, say, that wines from the appellation Volnay are judged to be lesser than those from the appellation Vosne-Romanée, with the latter home to six *grand cru* vineyards and Volnay none, when these communities are only separated by a few miles? Even more confounding, certainly to a non-Burgundian, how is it that the wine from Vosne's La Tâche could merit the highest rank, and thereby be worth more, than the wine from Aux Malconsorts, when these vineyards are separated by a few feet?

Indeed, within the classifications are a wide range of nuance and variety, which, as far as the vast, international cult of

Burgundy's devotees is concerned, is the result of much careful reflection, informed by the region's unique spirituality.

Ultimately, the INAO's methodology, as with all things in Burgundy, grows from that mix of science and mysticism—*terroir.* In order to fully appreciate the classifications for all that they are—and for all that they are not—one needs a bit of historical context and a willingness to take a leap of faith.

The INAO undertook their task having the blessing and curse of two thousand years' worth of attempts to define Burgundy. The Roman Empire was responsible for the earliest vines in Burgundy, which was then Pagus Arebrignus. In the last century before Christ, Virgil and Pliny were among the first to ponder why the wines from this region were so superb. In the first century AD, a Roman farmer, Columella, flirted with the notion of *terroir* when he wrote, "The vine planted in fat silt yields abundant wine, but inferior in quality."

Columella also may very well have been one of the first to recognize Pinot Noir vines planted in certain Burgundian soil was a magical combination. Experimenting with three different varietals, he determined that the grape that would be named Pinot Noir was "the best of the three... praiseworthy because it endures drought best of all, because it bears cold, if only it is free from rain... and especially because it alone gives a good name to even the poorest soil by reason of its own fertility."

In the fifth century, with the Roman Empire collapsing, the era of the feudal lords emerged in concert with the Benedictine monks. The Black Monks, as they were known because of their hooded robes, were formed in Italy in 529 by St. Benedict, who by all accounts didn't set out to birth an order. But as Charlemagne conquered and declared Catholicism the French religion, Benedictine monasteries sprouted throughout Europe.

The Benedictines adhered to a strict monastic life of work and prayer. "The Rule," St. Benedict had called it. In France, the monks' governing base, built in 910, was in Cluny, Saône-et-Loire, in western France. Constructed under the benevolent auspices of the Duke of Aquitaine, it was a grand edifice, emblematic of the coziness the Catholic Church enjoyed with the aristocracy. The French nobles fortified their alliance with the church with gifts of land.

In Burgundy, those gifts of land often didn't seem like they were good for much of anything. Columella's success with his vineyard, wherever exactly that had been in Pagus Arebrignus, undoubtedly took some doing, for the vine and the wine didn't come easy in Burgundy. The region then resembled prehistoric wilderness. The soil was riddled with bits of marl and limestone and made gritty with the atomized, residual remnants of the ancient ocean. All of it was covered by a forest of dense thickets, among jagged hillsides that dipped into lowlands prone to wind and frost.

Toward the end of the tenth century, the concept of *terroir* became ingrained in Burgundy's culture when two groups of holy men arrived. A band of exhausted, ragtag monks carrying a saint's corpse came from the west, and a more orderly group of monks traveled from the north, led by a rebelliously righteous former noble from Champagne.

The more disheveled bunch toting the body was the Order of St.-Vivant, twenty-eight monks whom the Normans had chased from their monastery near the Franco-German border. When these holy men fled, they grabbed only the important things: household utensils, liturgical supplies, and the body of Vivant, the fourth-century saint who had founded their clearly committed order.

Their decision to settle around Vosne was a fortuitous one.

They received the gift of an *abbaye* from a dying nobleman looking to buy away his sins. Count Manassès, persuaded by his wife, Ermengarde, the daughter of the former king of Burgundy, gave them an abbey on a hillside directly on the other side of the *côte* from Vosne. And none other than Hughes II, the duke of Burgundy himself, gave the St.-Vivant monks, so said the lease, all of his "uncultivated lands, woods and fields in Flagey and Vosne." Ironically, these St.-Vivant monks joined the Cluny, which was the order the other band of monks had just fled.

The other monks who had traveled to Burgundy followed Robert of Molesme. Molesme joined the Cluny sometime before 1060. A monk who held the Rule sacrosanct, he was horrified by what he saw at his order's headquarters: There was corruption and concubinage among the brothers. After unsuccessfully agitating for reform, and repeated disagreements with the Cluny, Molesme hit the road, taking twenty-one monks with him. Evidently judging that Burgundy was far enough removed from Cluny, he set up a camp in Bourgogne in 1098.

Molesme's order erected an abbey just to the east of the villages of Romanée, Flagey, and Vougeot. They christened it Cîteaux, after the reedlike grass. In their library, some of these Cistercians spent their days painstakingly, with quill and ink, making copies of scripture and other worthy works of the time. Other members of their order went out into the land that had been given to them, and put their backs into clearing the thickets and cultivating vines.

For monks dedicated to work and prayer, the gift of uncivilized land was indeed a godsend. The monks saw God's presence, his wonder and beauty, in all things. It was a deeply Catholic spirituality informed by ancient overtones of the Druidic philosophy of animism, which is the belief that all things possess a soul or living energy; and humans are but a piece of the natural phenomenon that God has provided, for His reasons, in His way, with divine purpose, all to be harnessed to serve Him, to honor Him.

Every patch of earth, then, every vineyard, with its specific soil and subsoil; with its particular altitude, pitch, and drainage; with its unique exposition to sunlight, wind, and rain, was singularly unique, as designed by God, and the vines there would produce a fruit and a wine like no other parcel.

With no commercial pressures, no pressure at all except to fulfill God's plan, to honor Him with their work and the fruits of that labor, the Cistercians planted and prayed, pruned and prayed, harvested and prayed, pressed and prayed, and then tasted—not what they had made, but what the Lord had given them the opportunity to birth. The more sublime the flavor, the more purely it conveyed the *terroir*, the more divine the wine.

To determine where one *terroir* ended and another ought to begin, to be able to maximize the natural wonder that God had created within each, the monks would savor and interpret the fruit before the harvest; they would taste and study the wines. According to legend, they knelt in the vineyards and tasted the earth. They would keep extensive records. They would adjust and refine.

For at least four centuries of harvests, they explored viticultural techniques and a small handful of varietals. They found that the wines from some communities that would be recognized as appellations were better than some others. They discovered that the composition of the land in the south was different from that in the north. In the south, the soil is heavy with marl, whereas in the north the earth is filled with limestone. In the south, the white Chardonnay grape grew better. In the north, Pinot was supreme.

They would learn that the most exalted, the most divine Pinot Noir came from a few parcels on an east-facing slope between the village of Vosne to just beyond the community of Chambolle. Equidistant between those villages, the monks built a spectacularly humble winery, Clos de Vougeot.

In time, certainly by the fourteenth century, when the first French pope relocated the papacy from Rome to Avignon and was serving Burgundy's finest wines, the nobles would get a taste of these wines and realize the profits to be had, and the nobles would want more of each. They began to tax the church off the best vineyards. The parcels the monks had deemed the best, on the slopes, which became known as the Slope of Gold, the nobles wanted those most of all. And they would get them.

Eventually there would be the legendary story of the Prince de Conti and King Louis XV's mistress, Madame de Pompadour, at dramatic odds, both casting an eye toward the very best parcel. Then revolution.

The INAO began their process by considering this sacred, ancient history, and then considered the most recent attempts experts had made to map the delimitations into something that made modern sense. There was no question that the definitive source was the work of Dr. Jules Lavalle.

In 1855, after many years traveling throughout all of Burgundy's appellations; after intense research in legal and historical archives, in wineries, in the vines, and tasting the wines, Lavalle published the *Histoire et Statistique de la Vigne et des Grands Vins de la Côte d'Or* (History and Facts of the Vines and the Greatest Vines of the Côte d'Or).

Among the facts Lavalle put forth were the most detailed vineyard maps and his ranking of the best appellations and vineyards. In descending order, Lavalle's rankings for the vineyards were: *hors ligne*, *tête de cuvée*, *1ère cuvée*, *2ème cuvée*, and *3ème cuvée.* On the maps, he crafted a color-coded key for each of his five categories. Lavalle's conclusions were consistent with those compiled by Jean-Alexandre Cavoleau, who published *Oenologie Française* in 1827, and Dr. Denis Morelot's 1831 *Statistique de la Vigne en Côte d'Or.*

Because Lavalle leaned on Morelot and Cavoleau, who in turn relied heavily on the records and opinions of the Cistercians, it seems fair to say that after such extensive "peer review" a thousand years later, when it came to selecting the best *terroir*, the monks knew what they were talking about. Lavalle's work, however, was the freshest, and mapped in detail. Therefore he endured as something of an expert witness on the matter of classification, which was exactly what the INAO needed.

The creation of the INAO and the push for classification were almost exclusively market driven. In the wake of the first phase of phylloxera, toward the end of the nineteenth century, burgundies were scarce and demand was high. The market was rife with fraud perpetrated by *négociants*. To fill orders, the agents were blending wines with no respect for the appellation: Wines from one village were mixed with another.

Wines from a prestigious vineyard got dumped into wines from another lesser vineyard. Unscrupulous *négociants* thought nothing of blending Pinot Noir with lesser varietals, in particular the Gamay. (In 1395, the Duke of Burgundy, Philippe the Bold, banned the growing of Gamay because this varietal of "horrible harshness," as he put it, was being planted in soil that could be used for tasty and more taxable Pinot Noir.)

Perhaps the most unconscionable act among the *négociants'* unconscionable acts, these agents went so far as to produce utterly fake wine, *vins artificiels*, in the form of flavored sugar water, which contained no fruit at all. They would sell that as a burgundy or if they found themselves short on more exquisite Pinot, they'd cut what they had with *vins artificiels*.

Onto bottles filled with this plonk the *négociants* slapped the typical labels with *négociant* brand, but also, often with the name of specific vineyards. This fraud flooded the market with terrible wine, which drove down the price, and thereby the profits, for the vigneron's legitimate wine. The *vins artificiels* business not only undercut the reputation of a vigneron's vineyard holdings, it devalued all of the wines of Burgundy. Many *négociants* were diminishing the long-term brand of the entire region for a short-term sell. Appellation, vineyard, varietal, *terroir*: None of it mattered. For the offending *négociants* there was nothing sacred; money was the holy grail.

When the INAO was finally mandated to intervene in 1935, it was establishing government-sanctioned classifications that would serve as both market standards and legal precedent. The *négociants* wanted as little regulation as possible; the growers desperately wanted government support. (This civil war did not go unnoticed by the likes of Schoonmaker and Wildman and, naturally, worked in their favor.)

The INAO took the perspective of the vignerons and created a defined class system: Regional wines can be produced anywhere in Burgundy. Village wines must originate from within the borders of a particular village. Wines from a specifically defined vineyard or *climat* with unique *terroir* are either *premiers crus* or the wines of the very finest *terroir* in Burgundy, *grands crus*. To misrepresent these wine classifications is a crime.

A cop's kid, a son of Burgundy, Inspector Pageault knew there had been and always would be plenty of grifters trying to move fake bottles. The crooked *négociants* of the early twentieth century weren't the first and they wouldn't be the last to commit wine fraud. Fake bottles of Domaine de la Romanée-Conti wines surfaced with regularity. As far as the police could determine, no one had a record of a crime like this one against the Domaine. Sure, vineyards and wineries have been vandalized here and there. But an act of agricultural terrorism against a vineyard, this type of extortion was unprecedented.

Inspector Pageault knew that fraudulent bottles of fine wines were a problem in that they could disturb the economy of the wine and wine auction market, and upset wealthy buyers who got ripped off. However, poisoned vines, poisoned *terroir*, this represented an attack on the sacred, unspoken communion between the vigneron and nature, between man and God. Inspector Pageault understood that this crime was an unprecedented attack on the spiritual core of Burgundy.

Pageault knew very well that *grands crus* were rare. Of the one hundred broad Burgundy appellations there were only thirty-three *grands crus*, which represented less than 2 percent of all French wines. Not surprisingly, six *grands crus* were clustered in the village appellation of Vosne. Two were in Chambolle. Whoever was behind this crime, Pageault suspected, could navigate a fine wine list, or at least knew enough to research Burgundy's complicated classification system. Pageault would soon learn that he was half right.

—

"There he is," Prignot radioed to her backup.

It was 8:40 p.m. when Leduc arrived home from his job at the kitchen supply shop. The thirty-nine-year-old had spent the day answering questions about cabinets and countertops.

"Pierre Leduc?" Prignot asked. She and her backup approached Leduc as he entered the front gate of his apartment.

"Yeah. What can I do for you?"

Prignot identified herself, showed the stunned Leduc the warrant, and ordered him to show them to his apartment. While her colleagues seized his computer and took a mouth-swab DNA sample, Prignot questioned him.

No, Leduc insisted, he did not send a package via Colissimo to a Monsieur de Villaine in Bouzeron.

Prignot asked him to explain why he tracked a package that was sent to Monsieur Aubert de Villaine in Bouzeron. "We know you did," she said. "That's how we found you. You tracked the package from an IP address we traced to this computer. So if you're lying..."

Leduc maintained he didn't know what she was talking about. He said he'd never heard of a Monsieur de Villaine. He supposed that Prignot was referring to the fact that when Leduc went to track a package he was having shipped to his apartment, the first time he tapped in the tracking code he had accidentally entered the wrong number. He said he had been waiting on a PlayStation he'd ordered.

It was right there. The packaging for the video game console was still next to the television set.

Within hours back at the Dijon branch, the crime lab's technical experts had scoured Leduc's hard drive. Nothing. The guy was telling the truth. Every case Pageault had worked on, he told Prignot, had one dead end, but he couldn't think of one with two. Bessanko and now Leduc.

As far as promising leads, the inspectors now had nothing.

Over the next week and a half, Pageault and Prignot developed no new leads. Just as Pageault began to take a hard look at the list of employees he had asked the DRC and Domaine de Vogüé to provide, on February 4 the DRC received a note addressed to Monsieur de Villaine. It was delivered via regular mail and instructed him to put one million euros in a brown paper bag, and the next evening, at midnight, leave it where he had found the two dead vines.

That the note arrived via regular mail struck police as odd. Perhaps the extortionist sensibly concluded police would be watching Colissimo—and they were. Maybe the extortionist realized that if he called Jean-Charles, as the police had tried to bait the extortionist into doing, police were prepared to trace the call—and they were, and not just on Jean-Charles's phone. Police had tapped just about every line into the Domaine and its senior staff.

But regular mail? Evidently the extortionist had great faith in the French postal service. Also confounding the police was the fact that whoever the architect of this was had directed that the ransom be paid in cash and delivered to a vineyard. Cash in a bag? Left in the open? Why not have the payment wired into an account? That was more along the lines of what Prignot had been expecting. Even if the police took the extortionist at his or her word, and whoever the pickup person was had no knowledge of the specifics of the poisoning—no knowledge of which vines had been attacked or how to prevent the death of more vines—certainly the police could follow that person.

Considering the extortionist had been so methodical and clever thus far, the strategy now being employed seemed uncharacteristically risky, bordering on stupid...or some kind of brilliance. Perhaps there was more going on here than met the eye, and this was all so deft that the police could not see through the plan. Maybe the extortionist was right now accomplishing his wish by rendering them all off balance and was off somewhere laughing at the idea of them scratching their heads.

Maybe the vines were not the mark and that threat was a misdirection for something else? Maybe the extortionist was not an extortionist and had a kidnapping planned? After all, whoever this was had discovered where Monsieur de Villaine lived. So, then, if a kidnapping was the ultimate goal, why not grab the monsieur there, or on his way there?

Maybe the endgame was something else entirely.

Now, not only was just about everyone a suspect, it seemed anything was possible.

With less than twenty-four hours to prepare, Commander Millet decided there was no way for his team to put into place the necessary strategy and logistics around Romanée-Conti, around Vosne, to ensure they successfully apprehended the bad guys and ended entirely whatever it was that was happening. At least not without the risk of tipping their hand. Millet figured the sting would require at least three dozen officers and hours of careful setup. So far the bad guys had been clever and careful, and the police didn't know where the criminals had their eyes and ears.

Millet wanted to try to buy more time and better circumstances. Rather than hastily attempt to arrange a sting and risk compromising their investigation and losing the criminals and the vineyard, now two vineyards, and maybe even risk losing lives, rather than deliver the money they would deliver a note.

Monsieur de Villaine was away on business. He had been reluctant to leave, but the police insisted. They advised that it would be best if he generally stuck to his routines and appointments, and continued to do whatever he would normally have done.

In the Grand Monsieur's absence, it was decided that Jean-Charles would write the letter. The fact that Monsieur de Villaine was away was actually helpful in establishing the credibility of the note.

The note said that Monsieur de Villaine was prepared to pay the million euros, but he didn't have that kind of cash on hand and would have to secure the clearance of representatives of the two families who owned the Domaine. The letter would beg for a bit more time, a few days was all that was required. Jean-Charles petitioned for patience and asked the perpetrator to call him on his cell phone with the time and location of the drop-off.

The next evening, promptly at midnight, Jean-Charles drove his Mercedes sedan out to the vineyard with Commander Millet hiding on the floor in the backseat. The two confirmed dead vines that had been removed were in the southwest corner of Romanée-Conti. Jean Charles took the road out of Vosne that ran along the south wall of the vineyard. The road twisted up the hill about one hundred yards and dead-ended at the woods.

As Jean-Charles and Millet had planned, when they reached the point in the road for the drop-off, Jean-Charles put the car in park and got out, leaving the car running. He walked around the car into the vineyard. He put the note at the base of a vine next to holes in the earth. There were no lights in the vineyards. Jean-Charles was reminded just how dark it can be out there.

Without lingering, he returned to the car and headed west on the road, up to the top of the hillside. Without going especially fast or slow, Jean-Charles reached the top of the hill and turned around. Descending the hill, the light from his Mercedes flashed over the spot where only minutes ago Jean-Charles had put the note. He told Millet he wasn't going to believe this, but the note was already gone. Jean-Charles was certain of it.

CHAPTER 12

Loss

Thirty-four-year-old Aubert de Villaine was driving as fast as he could toward Paris. In the passenger seat, his father, Henri, sat in a catatonic-like state, staring at the road straight ahead but seeing nothing at all. As much grief as Aubert was feeling, all he could think about was his father. Aubert knew Henri's heart was twisting. He didn't know what to say to his father. He wasn't sure there was anything that could be said.

And so, for the entire drive from Vosne to Paris neither of them spoke more than a few sentences. There was that kind of profound, pregnant silence that makes a long ride even longer. The rhythmic, soft thump-thumping sound of car tires over the cracks in the road and the soft whistle of the air current around the car became a kind of torture, all of it a maddening reminder of how far they had left to go.

The 1970s had started off wonderfully for Aubert: He and his wife were expecting a child. After his 1965 trip to America, he returned several times. In 1969, through the Wildman family, he met Pamela Fairbanks. One of Wildman's nieces and Pamela had gone to boarding school together. The Wildmans and mutual friends within the wine world had long been telling Pamela and Aubert that they should meet, that they would make a good match. Pamela was from Pasadena, California, the great granddaughter of Charles Fairbanks, who had served as a U.S. senator from Indiana and as vice president to Theodore Roosevelt, a man Aubert greatly admired. Pamela knew wine. She also knew France, very well, and loved it. She had spent her college years at the Sorbonne.

In 1969, Pamela and Aubert found themselves in New York at the same time. Aubert invited her on a date. One of the first things he noticed about Pamela was her smile and her laugh. Her spirit was light, such that it lifted his own. Explaining to Pamela that his mother was Russian, he took Pamela to dinner at the Russian Tea Room.

When Aubert came to pick her up for their date, Pamela was staying with an aunt. The minute Pamela opened the door and saw Aubert, she felt as if she'd always known him. As they walked the city together, she thought Aubert cut a very dashing figure. In an observation that would foreshadow their lives together, she also found it difficult to match his pace. She would always look back on her first dates with Aubert and remember herself thinking, *I don't know where this young man is going, but I sure would like to go with him.*

During their time together in New York and the many dates that followed—she would visit him in France and he would travel to America—Pamela was struck by his incredible *simplicité* and yet at the same time his *très cultivé* sensibilities. She quickly saw that his outwardly reserved façade was precisely that, a necessary defensive mechanism. The real Aubert was tender and loving. They married in 1971 in a Catholic wedding, but also in a civil union officiated by the mayor of Vosne-Romanée, Jeanine Gros.

Vosne being Vosne, Jeanine also happened to be the matriarch of the highly regarded Domaine Gros.

Pamela welcomed the move to rural Burgundy. She understood full well that Aubert was in line to become codirector of the Domaine, and she understood what that would entail. She was fully supportive, as long as he kept some distance and reserved time at home, just as his father and grandfather had done, for the two of them and for the family they looked forward to having.

The home in Bouzeron, she had agreed, would be a perfect place to raise children. Together, the newlyweds envisioned, they would raise their own vines and one day pass them on to their children. But what had started off as the very best news of incipient glad tidings ended in the stunning loss of miscarriage. Painful as it was, Aubert and Pamela held hands and reassured one another that these things happen. They would simply try again.

Meanwhile, together they planted a garden at their home in Bouzeron. They tended their *enfant* vines and expanded that family by acquiring more parcels around their home. Together, as Domaine A&P de Villaine, in 1973, they celebrated a different kind of firstborn, their first vintage, a Bouzeron Aligoté. Fittingly, the Aligoté is a white grape varietal that flourishes in trying circumstances where other varietals would not succeed, often at the bottom of the slope, where it can be very cold.

Aubert never imagined his circumstances would become more challenging. However, he had not yet reached the bottom of his slope. There was horrific news about Jean in Paris.

It happened in 1973. By then, Aubert had been working at the Domaine for eight years. During that time, he apprenticed in the *cuverie* under André Noblet, and he learned the management of the business from Monsieur Leroy and his own father, Henri.

Earlier that day, before Aubert and his father began their unexpected trip, they had arrived at the Domaine and were met by Madame Clin. She had stayed on after Monsieur Clin retired. The look on her face was unlike anything Aubert had ever seen in Madame Clin. Something was very wrong.

"Monsieur de Villaine," she had said to Henri, "you must call your daughter, Christine. Something has happened to Jean."

Christine, who was then twenty-three years old and Aubert's youngest sibling, shared an apartment in Paris with Aubert's youngest brother, twenty-eight-year-old Jean.

Henri dashed to the phone. Aubert watched his father listen to Christine and observed his father wither.

Christine informed their father that Jean was in the hospital. The night before, she and Jean had gone to a party at a friend's flat. Jean had complained of a headache and left early. On the way home, he collapsed. The doctors weren't sure, but the prevailing theory, and the one that would prove to be accurate, was that Jean had suffered an aneurysm.

Henri asked his daughter how serious it was.

"Papa," she said, "you need to come very quickly. They are doing what they can to keep him alive until you and Mama get here."

Aubert's mother, Hélène, and his sister, Marie-Hélène, traveled from their home in Moulins and met them at the hospital. Aubert's brother Patrick and their other sister, Cecile, would gather with the family later. The family joined around Jean's bed, where he was hooked to machines that showed a live heart but a dead mind.

Each of those there had time to be alone with Jean, to

kneel next to him and to pray, and to weep. Jean was very dear to Aubert. They were very close. Aubert admired Jean's sense of humor. They shared a love of poetry. Whenever the family would gather, everyone knew that Aubert and Jean inevitably would end up off together.

The medical staff unhooked Jean from the machines and he was gone.

While Aubert's mother would struggle for years to recover from her son's death, Aubert proved emotionally resilient. He took very seriously the Catholic teachings that assured him that he and Jean, and the child Aubert and Pamela had never met, would see one another in heaven. There, perhaps, as Aubert had planned, Jean would become the godfather to Aubert and Pamela's unborn child.

Over the next few years of the 1970s, very few people would know that Aubert and Pamela were enduring a second miscarriage, then a third, followed by the resigned acceptance that they would never have children of their own.

It would have been understandable if after three miscarriages Pamela and Aubert had felt more space between them. Instead, their bond intensified. They found family in each other. In this rocky *terroir* of life that God had provided, Aubert and Pamela tended to each other. Their love, like Aubert's faith, grew stronger.

It was about a year after Jean's death that the new era of the Domaine began. In 1974, Henri Leroy and Henri de Villaine left their roles as codirectors. It had been in the planning for quite some time. More and more, Leroy and de Villaine had been giving additional responsibilities to Lalou and Aubert. They would be the ones to inspect the account books. They would be the ones to host the buyers and guests who visited the Domaine. If they chose to, they had been empowered to hire new distributors—they had traveled around the world together to speak with potential distribution partners.

There was no ceremony, no party to commemorate the historic changeover. Aubert's father did not pass along any advice. Whatever wisdom Henri had to give, Aubert already had received. The one bit of advice Herni gave Aubert that Aubert often had found himself remembering was "*Le mieux est souvent l'ennemi du bien.*" The drive to make something perfect often ruins the good you have, or, as the Americans say, Leave well enough alone.

What there was to note the Domaine's new appointments was perfunctory. At the annual meeting of the two families, all of the stakeholders gathered in the Domaine as they did every December. The first order of business was to approve the new codirectors. Someone announced what everyone already knew, that the proposed successors were Lalou and Aubert. The stakeholders were asked to approve of the appointments by a majority vote. Along came the question, "All those in favor?" Every hand went up. That was that.

The change, and how it unfolded, in the wake of Jean's death, and while Aubert and Pamela privately endured their heartache, was the first time that Aubert fully comprehended one of the things his grandfather and father both had always said:

"The domaine is bigger than any one person or any one family. We will come and go, but the Domaine will live on. We are part of its history only for a short time. The Domaine is the Domaine."

As the new codirector of the Domaine, Aubert was about to make some of his first contributions to that history, by participating in an unprecedented wine-tasting competition, and with the release of his early vintages of DRC wines. Hard but valuable lessons would be learned.

One of the very first wine shops to carry Aubert and Pamela's first vintage of Domaine A&P Bouzeron Aligoté was a relatively new *magasin de vin* in Paris owned by a clever and charismatic British chap by the name of Steven Spurrier. Born into English privilege, from the time he was a small boy Spurrier was fascinated by wine. As a child he would rearrange the bottles in his parents' cellar over and over again, obsessing over the labels the way some American kids do with baseball cards.

Spurrier's love of wine drew him to Burgundy, where, in the early 1970s, he and his American wife, Bella, bought a home and became part of a clique of friends who would rise to be some of the most influential figures in Burgundy and in the wine world at large. Regulars included Jacques and Rosalind Seysses of Domaine Dujac; Aubert and Pamela; and recently transplanted Americans Becky and Bart Wasserman. (In a matter of years, Becky would launch an export business in Beaune and become something akin to the Godmother of Burgundy, but at that point, the couple were merely avid collectors and great company.)

In April 1971, Spurrier purchased an established but languishing Parisian wine shop, Caves de la Madeleine. Named after a nearby church, it was in a tiny arcade, the Cité Berryer, in a posh pocket between the first and eighth arrondissements. Eighteen months later, the shop next door became available and Spurrier picked that up, too.

Fluent in French and English and savvy, he converted the former locksmith shop into the Académie du Vin, which would offer wine education classes to his well-heeled English-speaking clientele. Spurrier had the expertise. If there was any doubt of that, right around that time, the *Revue du Vin de France* was giving a test for sommeliers; Spurrier waltzed in unscheduled, talked his way in, and was the only person to score a perfect 100 on the written exam. The moderator asked him to stay for the tasting component, but Spurrier declined, saying he had just wanted to see how he would fare on the written portion.

In 1976, with the Northern California wine scene coming alive, Spurrier hatched a plan to stage a tasting that would compare some of the finest wines of France with the some of the best American wines. He would swear that his intention was to host a comparison rather than a contest. He figured what better way to commemorate America's revolutionary independence (and not to mention market his relatively new business to the English-speaking wine world) than by gathering a knowledgeable panel and, among friends, assess the state of things. It was not, he would say again and again, his intent to give the revolutionary upstart winemakers in California a refined battlefield on which to engage the ancient establishment of French vignerons.

Spurrier held the noncontest on May 24, 1976, at the Inter-Continental hotel in Paris. There would be two rounds: one with reds, all of the Cabernet Sauvignon varietal; and one with whites: Chardonnay. There would be ten competing wines in each. The vintages would all be comparable. For the whites the range was 1971–73, and for the Cabernets it was 1969–73. While the range, to the untrained eye, might have appeared a bit wide, French wines are aged longer—two to three years—before they are released as a vintage, whereas California wines are cellared for only a year before they are put on the market. Admittedly, as far as fine wines go, what was being served was all very young.

Naturally, after much thought and outreach, Spurrier and

his staff assembled the most critically acclaimed wines of the time. As the best burgundies are Pinot Noir and Chardonnay, the white round was the only contest available to Burgundian domaines, and four were selected: Bâtard-Montrachet from the producer Râmonet-Prudhon, '73; Puligny-Montrachet, Les Pucelles, Domaine Leflaive, '72; Beaune Clos des Mouches, Joseph Drouhin, '73; and a Meursault Charmes, Roulot '73.

Himself included, Spurrier had arranged for a panel of eleven judges, nine of them French and with impressive credentials. Among them were Christian Vanneque, the sommelier of Tour d'Argent, the legendary Parisian restaurant with an equally legendary cellar and wine list; Pierre Bréjoux of the Institut National des Appellations d'Origine; Odette Kahn, editor of the *La Revue du Vin de France*; and Aubert de Villaine, the only Burgundian vigneron in the mix.

Spurrier regarded his friend Aubert as a logical choice since he had spent a fair amount of time in Northern California getting an education in the region's wine country from the likes of Dr. Winkler. Of course, as everyone in the French wine world knew, he also was the incipient codirector of Domaine de la Romanée-Conti, the greatest winery in all of France.

The contest began shortly after 3 p.m. and Spurrier explained what the judges assumed would be the case, that the tasting would be blind, meaning Spurrier's staff had gone through the typical protocol for blind tastings. They had removed the wines from their original bottles and would be serving them in neutral, uniform bottles—that task had been done an hour and a half earlier. Wines would be graded on a point system—0 to 20—based on the four criteria of eye, nose, mouth, and harmony.

Round one was the Chardonnays. Coming in at first place in Spurrier's noncompetition, with a score of 132 points, was an American, the Château Montelena, '73, with the Meursault Charmes in second with 126.5 points. The Clos des Mouches was fifth with 101 points; the Bâtard-Montrachet scored a 94 and came in seventh; the Puligny-Montrachet, with an 88, was eighth.

When Spurrier read aloud the results, although the French judges were reserved, the collective and universal reaction was shock. Spurrier was sure that round two would go differently. With California wines winning the first round, the French judges would believe they needed to pay more attention.

Round two was the Cabernet. Among the four Bordeaux in the tasting were a Château Mouton Rothschild, 1970, and a Château Haut-Brion of the same vintage. The round also went to a wine from California, a 1973 from Stag's Leap Wine Cellars, with 127.5 points.

Spurrier had tried to entice as many reporters as he could to cover the event but only one showed: George Taber of *Time* had decided to stroll over at the last minute. Taber wrote a story on the event, titled "Judgement of Paris," which ran on page 58 of the June 7, 1976, issue. *Time* was the most dominant newsmagazine in the United States, with a readership of more than twenty million. News outlets around the world picked up the story, including the *New York Times*.

Spurrier and his staff were annoyed that the press seemed intent on depicting the affair as a competition between French and American wines rather than a "discovery." The American coverage was all variations of the headline that ran in the *New Orleans Times-Picayune*: "California Wines Beat French Wines!" All over the United States, clerks in fine wine shops witnessed shoppers picking over French wines on their way to buying out their supplies of the California winners.

Word spread throughout France much faster. There were

immediate calls for Pierre Bréjoux to resign from his role as inspector general of the INAO. When he got back to the Tour d'Argent, the sommelier Vanneque was reprimanded. When Aubert de Villaine returned home, Lalou wasted no time and minced no words. She told him he had "spit in France's soup," and that he had just participated in a scandal that "had set the Domaine back a hundred years."

As far as the Grand Monsieur was concerned it was a tasting. Nothing more. A single event with only a single reporter present. The scores were the scores. But he didn't bother sharing any of that with his new codirector. He shrugged off her words. What was there to say? There was, he was learning, no talking to Lalou. He wasn't going to try to explain to her that he thought California wines were indeed becoming very interesting and that France should heed the "Judgement of Paris" as a wake-up call. The Domaine was about to get a wake-up call of its own.

That same year, in 1976, Aubert and Lalou traveled to California and, unannounced, walked into the offices of Wilson Daniels, a wine importing and marketing firm, in St. Helena. To use the terms "offices" and "firm" might be a stretch to describe the Wilson Daniels of that time. Win Wilson and Jack Daniels, who had met while doing their own business in the American wine market, had just formed the venture. Their offices were on the second floor of the feed and seed building. Visitors of the time, like Aubert and Lalou, came through front doors and found themselves walking across barn floors covered in seed and hay, and then up the steps.

Win was tall with dark hair; Jack, short and blond. Both preferred Hawaiian shirts to suits and had the easygoing personalities to match. It was their likable demeanor, along with a work ethic of hungry hustle, that had made them independently successful wine dealers and marketers, and now, together, doubly successful.

Aubert and Lalou introduced themselves, asked if Jack and Win had a few minutes—of course, they did—and told the partners that the Domaine was looking for a new distributor in the United States. Frederick Wildman & Sons had been sold to another distributor, Hiram Walker, a Canadian company specializing in spirits rather than wine. Aubert and Lalou agreed that it was time to move on.

Very shortly thereafter, the Domaine and Wilson Daniels were in business, with Wilson Daniels agreeing to annually pay up front for its allotment of DRC wines. The partnership began with Wilson Daniels picking up what was left of the 1976 inventory and began in full with the soon-to-be-released 1977. In total, it was an outlay of about $5 million for Wilson Daniels.

In the weeks leading up to the arrivals of the 1977 shipment, Win and Jack got calls from Lalou raving about the '77s. Really raving. Evidently, Jack and Win should prepare for wines with an especially fine finish. The way the American wine sales chain works is the winery sells to its distributors; in this case, Wilson Daniels was the exclusive distributor for the DRC. The distributor sells to a licensed wholesaler, and the wholesaler sells to the licensed retail shops, restaurants, and hotels. Not long after Wilson Daniels sold off its '77s, calls started to come in from customers that there was something wrong with the wine.

Clients were complaining of a fizzy characteristic that a still fine wine should not have and that often indicates that a second fermentation has occurred within the bottle. Fermentation is the winemaking stage when the grape juice transforms into the alcoholic beverage, as yeast converts the juice's sugar into ethanol and carbon dioxide. Vignerons like those at the Domaine rely on the natural or "ambient" yeast that comes with the fruit itself, while others may add yeast. It's one of many areas where man does influence the wine. Regardless, temperature and oxygen levels determine how long the fermentation takes.

At the Domaine, the fermentation begins within only a few short hours, or less, of the fruit being harvested. The grapes, at least at that time, were transported to the winery in crates stacked on a flatbed towed by tractor. The crates from each parcel were then dumped into massive wooden vats and the fermentation was under way. Fermentation can take anywhere from four days to two weeks.

For myriad reasons, the fermentation time for Romanée-Conti may differ from the fermentation time of La Tâche, and so on. When a second unintended fermentation occurs in the bottle, as was the case with the Wilson Daniels '77 DRC wines, it is because residual natural yeasts in the wine come to life. It is nothing harmful, but it is unpleasant and certainly not what anyone pays for.

Jack gave Aubert a call. At first Aubert did not believe it, but Jack assured him that, indeed, all of the wines were bad. Jack felt a little like a Rolls-Royce executive calling all the dealerships to inform them of a recall on the latest models. Lalou insisted that there was nothing wrong with the wine, maybe just a few bottles. Jack said no, he was quite sure the whole shipment was bad. Aubert disagreed with Lalou and insisted they trust Jack's judgment and allow Wilson Daniels to do what Aubert knew must be done to honor the customers and the Domaine's reputation. Wilson Daniels took the entire shipment out behind the building and drove figure eights over it with a tractor, and the Domaine returned all of the revenue it had derived from it.

That Aubert would concede that the Domaine's wines were unfit for American clients; that he would side with Jack and Win, put further strain on Lalou and Aubert's already tense relationship.

CHAPTER 13

Drop-off

Just before 7 p.m. on February 12, 2010, Jean-Charles Cuvelier pulled his Mercedes up to the front gate of the cemetery of Chambolle-Musigny. On the seat next to him was a canvas valise filled with one million counterfeit euros.

He held on to the steering wheel to keep his hands from shaking. His arms shook. He could feel his heart. He told himself to calm down, that if anything went wrong, he would be fine; the police were nearby and could be by his side within seconds.

On the previous day, the Domaine had received another letter via regular mail. The fourth and what would be the final piece of correspondence said that time was up. It instructed that the money be put in a cloth valise and be delivered to the cemetery immediately east of Chambolle-Musigny on February 12 at dusk. The note specified to leave the valise inside the cemetery, in a small ditch behind a row of bushes just to the right of the entrance.

Monsieur de Villaine was away again on business, in America, promoting the release of the most recent vintage of the

Domaine's wines. Jean-Charles called him and alerted him to the contents of the note and informed him the police had a plan.

By the time that note had arrived, Commander Millet was ready. He had a team briefed on what was then known of the case, ready to mobilize. The lead investigators had anticipated that the architect of this crime would not select the same site for the drop-off. They had correctly guessed the new spot would be within the area. Less than two hours after receiving the note, Millet had the full team of approximately fifty officers gathered in a room at the Police Nationale's Dijon office studying a map of the region.

There would be several teams of officers in small white vans, like the ones used by so many of the domaines. Careful, of course, not to draw attention, they would patrol the area around the cemetery bull's-eye. They would cover everything north-south, between Vosne and Chambolle; and east-west, between RN-74 to the small rural roads just on the other side of the forest-covered *côte*. There would be undercover officers on foot, wandering the two villages, and in fixed positions in the forest and tucked still among the vines.

The next day, the morning of the drop-off, Jean-Charles went to the Dijon station, where he was briefed. A transmitter had been stitched into the bag, concealed in the lining. Another tiny transmitter had been sandwiched in one of the stacks of bills. Jean-Charles was instructed to wear his Bluetooth headset. Before he approached the cemetery he was to turn it on; Francis Xavier, the hostage negotiator who had traveled from Paris, would be on the line. They would keep the call live but say nothing. If at any moment it appeared he might be in immediate danger, Jean-Charles was assured, the police would know it and they would swarm.

After the briefing, Jean-Charles thought about his *enfants.* He thought about calling his two daughters. Up to that point, as the police had asked him to, he had said nothing to them. The thought of doing so had not occurred to him until this moment. Now he found himself thinking of them intensely, wondering what they were doing, wondering what they would say if they knew.

His youngest, "Solène," was in her midtwenties. Jean-Charles was interested in names and their meanings, the provenance of them. Solène: First time he heard it was from a mother on a beach. Solène is a type of seashell. Jean-Charles thought the word was lovely. Solène grows from a Latin root that means "solemnly." His Solène was anything but solemn. She was ebullient. Everywhere she went, she was like a light, bright both in mind and spirit, eager for adventure. She worked in Paris with the fashion design company Chanel. Neither of his girls had ever given him any trouble, but Solène, his light, his sun sometimes put herself in a position where it rotated around her. She was like him. If Solène knew about all of this, of course she would be concerned, but, Jean-Charles told himself, she would likely understand why he felt the need to do it and she would have encouraged him to go for it.

His oldest, Raphaëlle, was in her late twenties—the ancient meaning of Raphaëlle is "God heals." Now, she was the solemn, serious one. Raphaëlle's dark eyes always pragmatically sizing up the world, vigilant for the worst-case scenario. Sharp, intolerant of injustice and willing to fight for her convictions, was it any wonder she became an attorney. She worked as a jurist for the French government and also lived in Paris.

Raphaëlle was protective of her younger sister and of their father, who sometimes forgot he was no longer an indestructible rugby player. Raphaëlle would have wanted to be walked through every detail of the police strategy, and once satisfied that the plan minimized all risk—convincing her of which would not have been easily accomplished—she, too, would have supported him, probably saying something like, "Papa, get these bastards." Raphaëlle was more like her mother.

His late wife. How he missed her. Annick. Means "gracious and merciful." Oh, and she was. Through it all. To the very end. It was impossible not to think of his girls, their girls, and not think of Annick.

She and Jean-Charles had met in high school. Jean-Charles would always hope that she fell for him, at least in part, because she thought he was handsome. Unlike now, with his belly, certainly then, at least, he was fit. Prime rugby shape. Annick would come to watch him play, and when she did, he lowered his shoulder and hit harder. Annick was an athlete herself. Fencing. How she could dance with that foil. He loved to watch her. A ballerina with a blade. Remarkable instincts—after all, Jean-Charles would joke, she did marry him. And she did not know what it meant to quit. Not even when the diagnosis came.

Breast cancer. When Annick heard the diagnosis, in 2001, she heard "*Allez!*" To her it was like a fencing match had begun. She quietly made up her mind that she would riposte, parry, advance. *Attaque au Fer!* She would win. And she did. There was the victory of remission, but then another match in 2005. It lasted for three years. She spent her final months too weak to get out of bed.

Jean-Charles had long been a fan of magic. During Annick's

cancer, he became more fascinated by it. He joined an amateur magicians' group. Once a month, he and his fellow magicians would meet and share the tricks behind their tricks. Yes, Jean-Charles turned more and more to magic as the science of medicine seemed to fail his wife. He would pray, but he would escape into magic. Annick loved that Jean-Charles loved magic. She loved that he believed in it.

When he would sit by Annick's bedside in those final months and weeks and days, first at home and then in the hospital, Jean-Charles would try to entertain her with a new trick. Something to make her smile. Something to make her see that what seemed so real and so certain was not. Annick liked the trick where Jean-Charles made an ace of spades disappear. Of course, the ace did not disappear. It was there all along. So was the cancer that took her life in 2008, after thirty-three years of marriage.

Jean-Charles continued going to the monthly magicians' gathering. He continued to believe in magic. That's part of what he loved about the Domaine. In every vintage, in every bottle there was magic. There was hope. There was rebirth. The magic of the Domaine, the magic of the family he had there, is what enabled him to get through Annick's long bout with cancer and then begin his life without her.

Jean-Charles thought it was magical that a high school teacher had met Aubert de Villaine and that the Grand Monsieur had invited him into that world of the DRC. He felt as if he owed a great deal to Monsieur de Villaine and to the Domaine, to defend that magic. The idea that someone would destroy it, poison it…Jean-Charles felt confident his girls and his wife would understand why he was so willing to carry that bag into the cemetery. In many ways, they were why he was doing it.

Jean-Charles decided it was best not to call Solène and

Raphaëlle. They would worry. And, as he thought of what the night would bring, he convinced himself there was nothing for them to worry about.

"You're doing great," the voice in his earpiece said. "Walk in, leave the bag, walk out. That's all there is to it. We're watching. We're right here."

Jean-Charles reached over, grabbed the bag, and set it on his lap. He took a breath. He opened the door and slowly walked through the cemetery's gates. Millet's theory about why the note had disappeared so quickly from Romanée-Conti that night was that someone had been there, waiting in the dark. Mindful of that, Jean-Charles's eyes scanned the tombstones as he walked in. Nothing. Only dead calm.

He left the valise right where the note had said to, then walked back to his car—he wanted so badly to run to it. He closed the door and drove off.

"Well done, Jean-Charles," the voice in his Bluetooth said. "Go home."

Less than forty-five minutes later, as Jean-Charles was walking into his home, about to pour a very full glass of wine and relax with his cat, Merlin, his cell phone rang. It was Inspector Pageault. "Jean-Charles," he said, "we got him."

Jean-Charles asked who the man was and was he working alone. Manu said he'd only had a few minutes with the man right at the arrest and that he expected to get more out of him at the station. That's where Pageault was headed now. So far, Pageault told Jean-Charles, the suspect had identified himself as Jacques Soltys, and so far Jacques Soltys was saying he'd been in prison before and he wasn't afraid of going back, and that he wasn't going to say any more.

CHAPTER 14

Secret Expeditions

Dusk, January 5, 1757. The French countryside where King Louis XV was visiting his daughter, Madame Victoire, was bitter cold and dark. The weather suited the political climate and the treacherous events that were about to unfold.

The Protestants in the south of France, led by Pastor Rabaut and even more so stoked by Jean-Louis Gibert, had become increasingly rebellious. King Louis XV's decision to dispatch musketeers to arrest his most vocal critics in the *parlement* alienated more of its members. Whether because of calculated politics or sincere compassion, more and more the magistrates were taking the side of the Huguenots. Protestant sympathizers in the *parlement* were led by the likes of Guillaume-François Joly de Fleury, who had suggested that the Catholic Church allow the Crown to at least acknowledge Protestant marriages as legal civil unions, without religious sanction. Versailles was not interested in the suggestion. Paris was rife with rumors that the Prince de Conti was covertly fermenting a spirit of revolution among the Protestants in the south.

King Louis kissed his daughter adieu just before 6 p.m. on

that January 5. He made his way through his entourage of guards and servants to his royal coach, which would transport him back into that tempest in Paris. As the king approached his carriage, a man pushed his way through the crowd. With one hand, he grabbed the king, and with the other he plunged a dagger into Louis XV's side.

Among the commotion of the gasps and screams, guards took the assassin into custody: Robert-François Damiens, a servant. Within the hour, royal doctors determined the king's wound was not a fatal one. Within the month, the trial was under way.

In the context of the tensions coursing through the Paris streets, the royal court—really, just about everyone—reasonably speculated that the *domestique* did not act alone in his attempted regicide. With no question of Damiens's guilt, the trial became more of an inquisition, run by a commission composed of more than sixty members the king deemed "loyal" councilors. One of the men Louis XV assigned to the panel was the Prince de Conti.

One way to interpret Conti's appointment to the panel is that it was evidence the king dismissed the rumors insinuating the worst of the prince, and that the king instead chose to believe the best of his cousin. It had always been that way.

Ever since they were children, Louis XV had held his older cousin in high regard. Even when Louis-François seemed to have stepped into the darkest of shadows, the king had seen the prince in only the most flattering light. Time and again, Louis-François proved that the king's trust in him was well placed.

As a teen, when Louis-François fatally shot one of his Jesuit tutors, Louis XV, who was then not much more than a child, believed it to be a terrible accident, a belief later confirmed by investigators. When Louis-François spent his way into a debt of 1.6 million francs, the king bailed him out, believing it was an investment in his cousin's true character, which shortly thereafter Louis-François seemed to demonstrate on the battlefield. Never mind the high point of Louis-François's military career occurred only after he had ignored one of his cousin-king's decrees.

In 1742, despite a royal order banning princes of the royal blood from taking part in the military, the prince attached himself to a French regiment that marched into battle in the Piedmont. But the king could not fault his cousin. A report Louis XV received of the battle was that "the only talk is of the brilliant success which the Prince de Conti has had there and which very far surpasses all hopes that been formed from it." Upon his return, the king toasted Louis-François, saying, "*de mon cousin le grand Conti.*"

In his role on the panel in the Damiens matter Conti impressed many as being the one who more than all the others was interested in pursuing the facts to wherever and whomever they might lead. During the two-months-long proceeding there came a point, on February 19, 1757, where there was a theory that the Jesuits might have played a role. Addressing his fellow judges, members of the royal court, and other spectators in and outside the chamber, Conti urged the panel to explore the possibility. Speaking directly to his fellow judges with great rhetorical flair, he said, "Remember that the judges would have terrible remorse if the criminal, at the point of death or during torture, reproached their inaction by indicating accomplices in an area where reason alone was saying to look for them. I would succumb to deep grief into which another assassination would take me, born of a principle I would have left unrevealed and still existing."

The panel voted against investigating the Jesuits, one of several moments when the prince described those dissenting judges as "slaves of the court."

In the court of public opinion, Conti's apparent pursuit of justice on the stage of the Damiens trial served to bolster his standing with the French *citoyens.* According to a newspaper account published in the final days of the hearings, Conti's intelligence and integrity were now as undeniable as the bravery and commitment he had demonstrated in war:

> Never has a law case merited more attention than this one of the miserable Damiens. Thus it should be investigated and pursued with all possible exactitude and activity. The Prince de Conti, whose superior talents in war are so well known, evidences in this affair the greatest knowledge of the laws, applying them with the greatest justice, concerns himself with the smallest details, and neglects no circumstance which might help discover the accomplices. Nothing escapes his sagacity. In all the meetings he has spoken with this noble and vigorous eloquence which the Roman Senate admired in Caesar.

Not quite three months after the attempted assassination, Damiens was executed by a method that had not been witnessed in France for more than 150 years. He was tied to a stake and set on fire. While still alive, breathing, though barely, his charred, melting body was thrown onto the ground and each of his four limbs were tied to horses that were ridden until Damiens quite literally and grotesquely was torn into pieces. The spectacle played out over two hours on a cold, damp night in Paris town center, before a crowd of horrified spectators.

If the Madame de Pompadour, who was then thirty-six years old, had had her way, the Prince de Conti would have been executed right alongside the treasonous domestic.

It would not be entirely accurate to say that Jeanne-Antoinette Poisson slept her way to the top; however, it would be revisionism to deny that it was the foundation of her strategy.

She was born on December 29, 1721, in Paris, into ignoble circumstances that would foreshadow her life. Her birth came with the rumor that she was the spawn of infidelity. Officially, as much as it can be described as such, her parents were Louise-Madeleine de La Motte and François Poisson. That her mother was Louise-Madeleine there is no doubt. As to François being the father, that was a subject of gossip and debate.

François's status was firmly middle-class. Shortly after he and Louise-Madeleine married, they moved into the Hôtel des Invalides, which served as a home and hospital for injured and impoverished war veterans. A bureaucrat assigned to the War Ministry, Monsieur Poisson was in charge of the invalids' food service, in particular, meat rations. In 1719 he was deployed to staff one of France's military engagements.

Louise-Madeleine was a gorgeous woman. A brunette with porcelainlike skin, according to one of her contemporaries, she was "one of the most beautiful women in Paris." Her beauty was matched by her cunning. Another record of Louise-Madeleine describes her as "clever as four devils." She had married François because he was the best option at the time, but she felt she deserved better than the commanding officer of a military *boucherie*, and she was never one to go without.

While François was deployed, Louise-Madeleine shared the

company of many men, noblemen mostly—dozens, according to the gossip of the times. When François returned and held his daughter at her baptism, few believed the child was in the arms of her biological father. Shortly thereafter, the couple had a son, Abel-François. A few years of normalcy, then François took off, fleeing charges of corruption and embezzlement linked to illegal speculating on wheat, all of it linked to a famine in Paris.

While Louise-Madeleine figured out what to do next, she sent five-year-old Jeanne-Antoinette off to school at the convent of Ursulines of Poissy. Five years later, Louise-Madeleine brought her daughter home and began to give her a wholly different curriculum than the one the good sisters could, or in good conscience would ever allow themselves to provide. In the words of Marcelle Tinayre, a neighbor and close friend of the Poissons, Madame Poisson gave her daughter the "education of a superior courtesan."

Louise-Madeleine saw a reflection of her life and beauty in her daughter. She believed that her daughter's natural endowments were the key to Jeanne-Antoinette securing a better life than the one Louise-Madeleine had had. Having experienced life with a lowlife butcher and then sampling the lifestyle of some of the noblemen she entertained, Louise-Madeleine determined that her little "Reinette," as she called Jeanne-Antoinette, could become, if she worked hard and put herself in the right circumstance, the mistress of the king.

As Tinayre would describe it in her eighteenth-century account, Louise-Madeleine gave her Jeanne-Antoinette "an education, which aims at enhancing all the seductiveness of a woman. The sciences, literature, music are turned to uses both of ornament and strategy. Madame Poisson knew the value of beauty, but she knew, too, that beauty, if it draws the man, does not suffice to hold him; that there are hours of melancholy or fatigue in which the fairest face is powerless, if not irradiated by an inward light; in a word, that man demands variety and that that woman will keep him who can satisfy this unconscious claim, native to the most faithful of his sex."

Jeanne-Antoinette learned the harpsichord, dance, and elocution. She and her mother talked openly about putting her in a place to catch the king's eye, to scale "the Olympus of Versailles."

They networked her into the circuit of refined gatherings in Parisian drawing rooms. Along the way, she was spotted by a banker, Charles-Guillaume Lenormant.

In March 1741, at the age of twenty, she married Monsieur Lenormant. He had a country home in Étoiles, a house in Paris, an impressive salary, and most important, his was a social circle that overlapped with royalty, a world where ladies would take coaches out to watch the gentlemen and nobles hunt foxes in the Forest of Sénart. It was at one of these hunts, in 1744, three years into her marriage, that Jeanne-Antoinette at last caught Louis XV's eye.

Reinette made sure her coach was positioned in his eyeline. As Tinayre recorded it:

> *At that moment the youthful beauty of Mme d'Étoiles was in its full perfection. Life had not yet laid a finger on the fragile bloom of complexion where even the shadows were pearly, the complexion of pure blonde, a* déjeuner de soleil *[feast of sunshine], which betrayed a lymphatic and passive temperament, but in its early springtime made one think of all of the fairest, frailest marvels—the opalescence of a shell, the rosy heart of woodbine... the brow was made to have the hair thus drawn straight back, then lifted in soft waves obedient to the head's pure*

> *line—that chestnut hair. The eyebrows were two fair, unbroken arches, in the eyes whose hue was ever changing. Were they blue, or green or brown?...Lissome was the form...sufficed to fill a manly hand—and all this charming personality in its spreading skirts of puffed brocade, its ribboned bodices afroth with lace, its dainty slippers, with its little knot of flowers on shoulder or on bosom seemed to draw the line between the last degree of elegance and the first of aristocracy.*

How could Louis XV not notice this vision? When his carriage passed by, his eyes found her and lasciviously lingered. He was thirty-four, and here was this twenty-three-year-old flower, so clearly in full bloom. Their connection was so immediately apparent that over in the carriage that held his current mistress, the Duchesse de Châteauroux, and her friend, the Madame Chevreuse, they could not help but notice. Chevreuse remarked that the "little d'Étoiles woman was looking even prettier than usual." In response, right there in the carriage, the Duchesse Châteauroux kicked her friend until Chevreuse was unconscious.

Within weeks the king arranged for the butcher's daughter and her husband to attend a costume party at Versailles. That evening he arranged for the husband and wife to be separated and for the wife to join him in his private chamber. Jeanne-Antoinette, more or less, never left. She was ensconced at Versailles. The king compensated the banker for the returns on the investment that he had made in his wife.

Aware that a name change would be required to erase the awkward residue of her now former common life, Jeanne-Antoinette changed her title to the Marquise de Pompadour. Within the year, in May 1745, according to the protocol necessary to be recognized as a member of the royal court, Madame de

Pompadour was presented to the court, and more specifically, to Louis's wife, the queen.

Witnesses of the historic meeting would recall that the Madame de Pompadour greeted the forty-two-year-old queen with, "Madame, it is my most ardent desire to please you." The queen was said to have later remarked that their conversation had been a "long one of twelve sentences." From that point onward, it is unlikely they ever spoke more than a dozen lines again.

One of the royal ladies of the court who escorted the Madame de Pompadour to her audience with the queen that day was the Princess de Conti. The princess had been among the first to welcome her into the court. Yet her husband, the prince, had no interest in stroking the latest pussycat to hop onto his cousin's lap. The Madame de Pompadour tried to win Conti's favor, but none of her charm, no flattery, warmed him to her.

Members of the court noted he said very little to her, and when the prince did speak to Pompadour, he made clear what he thought of her; that no matter what her title, she was not a noblewoman. During a discussion that he and the king were having regarding the tensions with the Protestants, for which the Madame and her handmaid happened to be present, Pompadour said, "Do you never lie, Prince de Conti?" The prince responded, "Only on occasion... to ladies."

In 1752, King Louis made a surprise announcement: He was going to bestow upon Madame de Pompadour the rank of duchess. The optics of the honorific meant that Pompadour continued to have a special place in the king's heart. Pompadour was now able to sit at public supper and ceremonies with the king and queen. The king, it seemed, also hoped that the outwardly exalted title might soften the realization for Pompadour that she was no longer his favorite in bed. His new favorite was at the very ceremony where Pompadour was made a duchess, but soon she, too, would be gone, with many, many more to follow.

From her window at the palace at Versailles, the newly appointed duchess watched them shuttled to a discreet entrance to the middle-aged king's private chambers in ornate sedans carried by footmen and in carriages. *Petits oiseaux*, as they were known around the palace. Small birds. Being brought to his birdcage. To flutter about and land on him as she once did.

Madame Pompadour embarked on a new strategy to maintain her proximity to the king, by serving as adviser on affairs foreign and domestic. As long as she could chirp in his ear and he regarded her as indispensable, she would have respect and security. There was the training that her mother, clever as the four foxes, had given her: "Beauty, if it draws the man, does not suffice to hold him...."

Pompadour disagreed with every bit of advice that Conti gave the king. Where the prince would advise the king to compromise with *parlement* and the Jansenists and the Protestants, she would remind the malleable and melancholy Louis XV that he was still the king and this was still his monarchy and that he should hold firm to the royal decrees made by him and his predecessors. She was supportive, if not a driving force, behind the king's decision in 1755 to send in the musketeers and quash the parliamentary dissension with force.

She wasn't in an especially merciful mood. The previous June of that year her own daughter, ten-year-old Alexandrine, had died of peritonitis off in a convent. Pompadour was not able to get away, but her ex-husband was with their daughter as she died. But there was more to her reasoning than a heart hardened by her daughter's death and the spectacle of girls not much older than her daughter being carted into the king's bedroom.

Pompadour had worked hard to earn her place among the aristocracy, alongside the king. She had lately taken on the role of arranging for the illicit births of the king's bastard children to occur as if they never happened. She wasn't about to allow herself to be knocked off the summit of Olympus by a smug prince.

At every turn, she contradicted the prince and blocked the appointment of his friends and even his own chances for promotion. Toward the end of 1756 she supported the king's decision not to put Conti in command of a French army near Prussia and she fully supported the king's decision to strip Conti of his role overseeing the Secret du Roi and thereby undermine the prince's chance to be installed as the king of Poland. Both were promises the king had made to the prince. Conti was furious.

"Because I have not given him the command, which in all likelihood will assemble on the lower Rhine," the king remarked, "he says he is dishonored. This is a word one puts forward constantly nowadays, and which shocks me infinitely"—"shocks" meant to be received slathered in sarcasm. "Perhaps he will put some water in his wine."

The hope that the public and all of the different political and religious factions had held when Conti was advising the king was quickly evaporating. Now, in the words of one official, "public discontent is everywhere. The estrangement and bullheadedness of *parlement* has become stronger than ever. Those who were the most disposed to submit have retracted. One talks only of murder and poisoning. . . . In short, there are nothing but complaints and murmurs and protests against the ministry. It is about Madame de Pompadour they complain the most."

Then, with the start of the new year, on January 7, 1757,

came Damiens with his dagger. Within days of the savage show of the Damiens execution, a royal edict was declared, proclaiming death to "all those convicted of having written or printed any works intended to attack religion, to assail the royal authority, or to disturb the order and tranquility of the realm." It was a decree that sounded as if it were dictated by the Duchess Pompadour herself.

Another explanation for why the king might have installed his cousin on the panel investigating Damiens may have been that in the event his mistress was correct in her suspicion that the prince was not to be trusted, Pompadour might at last have prevailed upon the king to keep the potential enemy close while she dispatched the lieutenant of the French police, Nicolas-René Berryer, and an agent with Protestant credentials now hired for the task, Jean-Frédéric Herrenschwand, to investigate the rumors of the prince's involvement in the Protestant uprising allegedly brewing rapidly in the Bordeaux region.

In the early months of 1757, as that investigation began during the Damiens, Herrenschwand gained the trust of Jean-Louis Le Cointe, the Protestants' chief diplomatic representative in Paris. It had been Le Cointe who had helped facilitate the meetings between Conti and Pastor Rabaut in the abandoned hotel along the waterfront. Herrenschwand got Le Cointe to confide in him.

Le Cointe revealed that he and Conti met almost daily and were in communication with English agents who had been sent to the region in that early part of 1757. According to Le Cointe, Conti had sent Rabaut a memoir outlining questions and plans for a revolt. Allegedly, Conti had told Le Cointe, "It is not uniquely for the people that I act, but also for myself. For the moment that there are no longer any laws, the throne would be for the first occupant, and I have interest where the rights of my family are concerned."

Le Cointe said that many in the south prepared to take up arms; that there had been a Protestant synod where a handful of pastors, Gibert and others, had committed they "were prepared to give the prince carte blanche."

By June 1757, Herrenschwand had insinuated himself directly with Pastor Gibert: "I found him to be resolved and unshakable in his ideas. This miserable person even dared to tell me that he would defend them and that he was in a position to do so, having about four thousand well-drilled men who lack neither chiefs nor arms, and that if they were not successful, in his design, then he would decide to leave the kingdom with about fifteen thousand *religionnaires*; that all the arrangements for this departure had been taken beginning at the present. He defied whoever it might be to stop it given the measure by which it was assured." The spy noted that Gibert never admitted direct contact with Conti.

Intelligence Herrenschwand gathered indicated that whatever correspondence Conti had sent to Rabaut, the pastor had destroyed, and for his part, Rabaut appeared uninterested in an armed revolt.

By mid- to late summer of 1757, Herrenschwand filed an ominous and what would prove prophetic report, which he sent to Paris. He stated that he was worried that Gibert had the ability to muster twenty thousand men with arms, and with help of an invading force, that would be difficult to suppress.

"I hardly dare say what I think," the spy wrote before saying what he thought. "But there is every reason to believe, that the person in question has already made attempts on the exterior as he has made in the interior; many reasons make me suspect this. I hope

God wills that I am mistaken, but it is of the greatest consequence not to lose sight of this affair; of all the objects which should occupy the Government, this one seems to me to merit the most attention.

In separate correspondence he wrote in August, "That the Prince risked nothing; that no one could ever prove anything against him; that he had everything in order as much at home as abroad.... Ministers seemed to take pride in the fact that sooner or later the prince was going to change religion and in consequence unite all Protestants."

On September 20, at the dinner hour, 135 sails of the British fleet were spotted off the coast of France. Eighteen warships, six frigates, two hospital ships. The fleet was observed near the Île-de-Ré, traveling south. Among the armada was the *Royal George* with one hundred cannon; the *Ramilies*, the *Neptune*, and the *Namur*, with a total of ninety cannon. British marines aboard totaled at least eight thousand. About 6 p.m. on that September twentieth, the fleet anchored just at the entrance to the Bay of Biscay, under the command of Jean-Louis Ligonier.

Ligonier had been raised in a French Protestant home. His family had left France when he was a boy because of the tyranny enacted upon them. It was Ligonier who had recommended to William Pitt, England's minister of war, that the fleet approach France through Rochefort, just north of Bordeaux. Ligonier had solid intelligence that the port was vulnerable. He also had word from Gibert that he and his men were ready and had the support of a powerful prince.

Initially, the British assault could not have gone better for them or worse for the French. After having stayed anchored just outside the entrance to the bay throughout the twenty-first and the day of the twenty-second, that evening one portion of the fleet traveled into Biscay and anchored; then another group sailed in and anchored.

At 8 a.m. on the twenty-third, about eight ships leapfrogged from the rear directly toward Rochefort, stopping short of the Île-d'Aix, a small island fortress midway between the entrance of the bay and the mainland of Rochefort. Aix responded with only three hours of cannon fire before it surrendered. For the next six days the British fleet took prisoners at Aix and burned its military installations to the ground. With Rochefort's outer defenses neutralized, Ligonier's easy winning of the coast seemed guaranteed. But the fleet would never advance farther. On October 1, Ligonier pivoted his ships and their sails disappeared over the horizon.

In the weeks and months that followed, there was much introspection and postbattle analysis in Paris and London. The officials and citizens in each country wanted explanations. In England, the British demanded to know why their "elephant had labored to give birth to a mouse." In France, the tenor of the inquiries came with pride, tempered by Versailles's wish that not too many questions be asked.

For the French, Herrenschwand had proved to be the hero. Based upon the intelligence he had provided during those summer months, King Louis XV had dispatched six battalions of guards into the towns Gibert had indicated were inclined to rise up. The troops spent weeks confiscating arms and took up quarters in many of the homes. The French had taken the precautionary measures to snuff out the sparks before they could light something more explosive. The moves worked, though they might not have if the British fleet had not been delayed.

The "Secret Expedition" took time to get King George II's approval. In fairness to the king, he was facing a cat's cradle of military engagements in the region and this campaign, if pursued, would affect nearly all of them. The king was receiving petitions for assistance from his ally, King Frederick II in Prussia, that he was running out of hands to fend off the Swedes and Russians. If the English didn't help, Frederick said, he'd have no choice but to seek an alliance with the French. Not to mention Her Majesty's troubles with the upstarts in the American colonies.

By the time the British fleet left Portsmouth, England, on September 8, they were at least three weeks behind the schedule of the plans and timing originally set for in the schemes for the Secret Expedition. When the English fleet attacked the Île-d'Aix and in all the time before and the short time after they were in the bay, the English saw no sign of the promised native insurgence. Ligonier decided to abort. Because all involved in London were embarrassed by their "mouse" that barely squeaked, they let the matter fade as quickly and quietly as possible.

Louis XV accepted that Herrenschwand's intelligence implicating the Prince de Conti in the Secret Expedition was accurate. But he could not act on it. If it were made public that the wildly popular prince had lost faith in his cousin-king and orchestrated such an attack, the king feared it might expose the frailty of his fraying monarchy. The people and *parlement* might be encouraged to act so boldly.

Of course, the king, being the king, could have arranged for his cousin to vanish. It would seem even still Louis's affections and admiration for the prince endured. There is the notion of him trailing Pompadour through Versailles making the prince's case:... Herrenschwand's evidence against Louis-François is circumstantial.... The spy himself had said nothing could be proven against the prince.

For reasons political, but also profoundly personal, the exasperated king, who talked openly and often of his desire to walk away from the crown, was relieved to merely be no longer troubled by the prince and whatever it was that he had or had not done.

After the debacle of the "Secret Invasion," Pompadour at last had another wish come true: The Prince de Conti was not seen at court and otherwise kept the lowest of profiles. He spent time at the Palais du Temple and his private residences: the Hôtel de Conti in Paris and his retreat at L'Île-Adam, well south of the city. L'Île-Adam happened to be a Protestant stronghold, not that that necessarily reveals anything about Conti's religious leaning. As with all things related to the prince, it was never clear what he believed, or if he believed in any god at all. One of the few times Louis-François did surface in the public was to purchase La Romanée, but even then he was a figure in the shadows.

In the sixteenth century, the Cluny monks of the Order of St.-Vivant and the Benedictines at Cîteaux learned that when the ancestors of the dukes and nobles of Burgundy gave those gifts of land, there were strings attached. Or so the aristocracy was now informing them by way of steep taxes. The monks of St.-Vivant were left with no choice but to auction some of their holdings, including their vineyard, Creux des Clos, a small parcel of vines located just west of Vosne, at the base of the hillside.

Over the next century, Creux des Clos was sold several times over, purchased in 1631 by Philippe de Croonembourg, a military captain and nobleman from Flanders. Monsieur Croonembourg was the one who, sometime before 1651, renamed the vineyard La Romanée. The provenance of the name would forever remain a mystery. One possible explanation is also the simplest: Philippe was inspired by the ancient French term *Romanie*, which was often invoked throughout the fifteenth and sixteenth centuries to describe a very fine claret.

Regardless of his exact reasoning, if Philippe had rebranded the vineyard because his instinct was that the name Creux des Clos lacked a certain marketable *je ne sais quoi* and believed that La Romanée conveyed an alluring romanticism that would entice customers, he chose well.

Until the last half of the 1600s, wines from the Côte de Beaune were regarded as superior to those from the Côte de Nuits. Croonembourg's wine changed that. By 1733, La Romanée was valued at five times that of the next most prized wine from anywhere in the Côte. Between 1750 and 1760, a single *Romanée queue* (an old French wine term meaning a quantity of 456 liters) was worth between 1,200 and 1,400 *livres* depending on the quality of the vintage. The Croonembourgs shrewdly sold La Romanée only in *feuillettes* (a quarter *queue*), making it more rare and more expensive. By way of comparison, the wines of Clos de Vougeout were going for 200 *livres* per *queue*.

Despite the fact that La Romanée was far and away the priciest wine in France, when André Croonembourg, the fourth generation of his family to own the vineyard, died in 1745, his wife and children were left in debt and forced to auction off the prized vineyard.

The prince, whose ancestral roots were in Burgundy, with eyes and ears all over, would have been among the first to know that the region's most prestigious vineyard was for sale. Not only did he want this vineyard, judging from the covert steps he took to acquire the parcel, he ached for it. The prince pursued La Romanée as if it were the target of a *Secret du Roi* mission.

Because of his noble ancestry in Burgundy, the Prince de Conti was a lord of Nuit and d'Argilly. His family had received that land from Louis XIII in 1631 and owned vast domains in Côte de Nuits. Logically, if the prince wanted to buy this vineyard in the heart of Burgundy, he could have announced himself as the buyer and almost certainly his name and wealth would have chased off any other interested buyers. Yet the Prince de Conti himself did not bid on La Romanée. He dispatched a proxy to pose as the interested buyer. While buying through a straw man wasn't especially unusual for the times, the fact that the prince chose this route is curious, made all the more intriguing by the proxy he chose.

For his collaborator in this endeavor, the prince enlisted Jean-François Joly de Fleury, who was the intendant of Burgundy, akin to the governor. Joly de Fleury's father had been one of Conti's most outspoken allies on Protestant matters at *parlement* (and perhaps, even, a trusted aide in arranging the Secret Expedition). The Joly de Fleury family were no fans of Pompadour.

In addition to the covert methodology the prince employed to make the bid, the effort to acquire the vineyard was made even more interesting by what he offered to pay for it. The going rate for an *ouvrée* (about 4.2 acres) of vines was then about 200 *livres*, but Conti authorized Joly de Fleury spend 92,400 *livres*, or 2,310 *livres* per *ouvrée*—or more than eleven times the going rate!

No matter how much the prince loved good wine, and he did; no matter how badly he wanted this vineyard, clearly he did; no matter how wealthy he was, that Conti paid such an astronomical price is vexing. Especially considering what he did with it after he purchased it.

The Croonembourg family had managed to build La Romanée into a brand that had become the most expensive and highly coveted wine in Burgundy, which made it one of the most expensive wines, if not the most expensive, in France. With the network and financial largesse of the prince, Conti could have grown a fortune from his fortune. At the very least, he could have earned back what he'd spent on the parcel.

The prince continued to pour money into La Romanée. He spared no expense on the viticulture. He would visit Vosne and meet with his vineyard manager, stressing quality over quantity; the prince instructed him to do whatever was necessary to grow the best grapes for the finest wine. Yet none of his investment in the vineyard was for the sake of commerce. The prince removed La Romanée from the market completely and cellared it all as his own private reserve.

Why the secrecy? Why pay such a price?

One plausible explanation for his secret expedition into Vosne might be Pompadour. By 1760, Madame de Pompadour was thirty-nine years old. She was, without question, one of the most influential figures at court. The duchess had ensured her trusted confidants were elevated, including Berryer. The former head of the French police was made a member of the king's cabinet, where he continued to oversee the police and spy networks, along with greatly expanded powers.

It would have been extraordinarily unlike Pompadour to simply let her nemesis, the prince, waltz off unmonitored. After all, in her eyes, the prince had orchestrated an attempted assassination on the king and almost single-handedly triggered an English invasion of France that would have overthrown Louis—not to mention jeopardize her own well-being. With the prince still out there, she would have wanted to know his activities.

Even in the unlikely event the duchess did not have Berryer keep the prince under surveillance, it would have been unlike the spymaster Conti to assume he was not under Pompadour's many watchful eyes. If the Madame were to have learned that Conti wanted this prized vineyard, she may very well have bought it out from under him, just for spite. Thus the prince would have enlisted a front man, and someone whom he knew he could trust. Considering that Joly de Fleury had been Conti's ally in Paris, his son Jean-François made for a fine choice. The family Joly de Fleury would have relished any opportunity to thwart Pompadour. Similarly, it would make sense to make an offer that would expeditiously seal the deal.

But why pull La Romanée off the market? Why not flaunt it after you've got it? Maybe even send a case to Versailles addressed to the madame with a note written in pinpricks that says something like: Share it with the ladies. Sincerely, the Prince de Conti.

Why pull it from the market when all of the nobles, the bishop of Avignon included, wanted it and were willing to pay for it? Perhaps precisely because all of the nobles, the Bishop of Avignon included, wanted it and were willing to pay for it.

While the prince kept a low profile, that did not mean he ceased being the prince and stoking the flames of insurrection already burning with his countrymen. The guests that he and his mistress the Countess Boufflers hosted were often a who's who of the French Enlightenment. In fact, some of them were the architects of the Enlightenment. Rousseau, Voltaire, and Mozart were friends and guests of the prince.

In 1754, when Rousseau had published his essay, "What Is the Origin of Inequality Among Men? And Is It Authorized by Natural Law?" he had put himself at odds with the French establishment. In his subsequent works he continued to dare to promote civil rights and freedom of religion, thought, and speech. Conti and his friend the Duc de Luxembourg were patrons of Rousseau because they shared his views, but also because the more enlightened the *citoyens* became, the more it eroded the power of the monarchy of Louis XV and Pompadour. The archbishop of Paris condemned Rousseau and burned his works, which ultimately earned a warrant for his arrest and forced him to flee Paris. For a time Rousseau took shelter with the prince.

When Pompadour first arrived at Versailles, she celebrated and supported Voltaire's work. However, as his plays and essays and opinions became more critical of the Catholic Church and the monarchy, their friendship ended. Voltaire celebrated Britain's progressive embrace of freedom of speech and religion, and Shakespeare. Most notably, he channeled his observations into a series of essays called *Lettres Philosophiques sur les Anglais* (Philosophical Letters on the English). Exiled and often on the run from French authorities, Voltaire, too, found safe harbor with the prince.

And then there was Mozart. From the time Wolfgang Amadeus was a boy, the Contis had recognized his talents and praised and supported him. As a child, Mozart would visit the prince, play for private gatherings at the Contis' home, and listen to the politically fueled conversation. Such visits almost certainly shaped Mozart's work. Within a matter of years, with France on the brink of revolution, Mozart would compose the comedic opera of Beaumarchais's *The Marriage of Figaro*.

On the surface, the opera was satirical, but it masked subversive political commentary. In the first act, the servant Figaro dares to challenge his master. Figaro is an intelligent man of integrity, while the aristocrats are cruel, self-indulgent fools. Louis XV's successor, King Louis XVI, attempted to have the play banned. Despite the king's efforts, the show would go on in Paris, causing a riot in the audience in which three people were trampled to death. Napoleon would call it "the French Revolution on the stage."

Voltaire, Rousseau, Mozart, they would gather around the prince, the man at the center of everything and nothing—the man who had just bought La Romanée and put it beyond the reach of the monarchy. The prince's wine, much like his independence, could not be bought by the crown or the aristocracy, but the Prince de Conti was happy to give as much of the wine to these men as they liked. Louis-François was pouring a bit of his own subversive La Romanée revolution into their glasses.

The Prince de Conti died at the age of fifty-eight from what was probably cancer. The year was 1776, the year America secured its independence, France's revolution to begin thirteen years later. From the Prince de Conti's son, the New Regime would confiscate property, including La Romanée and its winery in Vosne, and put them up for national auction.

The advertisement for the 1794 auction would read:

> *La Romanée. Is a parcel of vines famed for the exquisite quality of its wine. Its situation in the vineyard territory of Vosne is the most advantageous for the perfect ripening of the grapes; higher to the occident than to the orient, it receives the first rays of sun in all seasons, being thus imbued with the impetus of the gentlest heat of the day.... We cannot disguise the fact that the wine of La Romanée is the most excellent of all those of the Côte d'Or and even of all the vineyards of the French Republic: weather permitting, this wine always distinguishes itself from those of the other* climats *of predilection, its brilliant and velvety color, its ardor and its scent charm all the senses; it is then balm for the elderly, the feeble and the disabled and will restore life to the dying.*

Toward the end of the promotional copy there was this phrase: "...jealously coveted by La Pompadour who failed to succeed in her intrigues."

CHAPTER 15

Quelle Pagaille

On September 1, 1988, the French minister of agriculture held a press conference at the ministry's Hôtel de Villeroy headquarters in Paris to deal with a controversy suddenly swirling around the Domaine de la Romanée-Conti. Minister Henri Nallet felt he had no choice but to formally address the media that Thursday, as it had become clear to him, to nearly everyone involved, that it was necessary to set the record straight. Things had gotten out of hand. Over the previous few days the whole affair had become, as the *New York Times* would report it, a stumbling block for "Franco-Japanese relations."

Quelle pagaille.

What a mess.

Standing before the not-so-small small group of journalists from around the world, Nallet said, "Romanée-Conti is like a cathedral. There is no question of letting a part of France's cultural patrimony get away." Nallet went on to say that Romanée-Conti should be regarded as a precious "work of art," and the French government has the right to prevent it from falling into foreign hands.

The story the *Times* published on the matter at hand came

with the headline "Wine Plans of Japanese Upset French." A spokesperson for a very frustrated Monsieur Aubert de Villaine was quoted as saying, "We have no intention of selling the vineyard." There was a comment from an exasperated Norio Ushiyama, the president of Japan's department store chain Takashimaya: "It appears that the French press has tried to undermine us by saying that we want to buy Romanée-Conti."

Under the terms of the deal Monsieur Leroy had brokered in 1942 when he bought half of the Domaine, his company, Société Leroy, secured the right to be the exclusive distributor for DRC's wines everywhere in the world except for the United States and Europe; those markets remained with the de Villaines. Since 1970, Takashimaya had worked with Société Leroy to market the Domaine wines, along with Société Leroy's own wines, in Japan.

In August 1988, representatives of Takashimaya and Lalou Bize-Leroy, who was now at the helm of her family's company, struck a deal in which Takashimaya would buy 33.6 percent of Société Leroy for $14.6 million. The way Ushiyama explained it in the media, Takashimaya's investment in Leroy was a natural evolution of their long-standing partnership. "We want to help expand the sale of French wines in Japan," he said, and Lalou wanted to focus more on the quality of the wines.

Japan, along with the rest of Asia, was emerging as one of the very best markets in the world for fine French wine, which the vignerons welcomed. On the other hand, the French weren't so crazy about the fact that the Japanese were buying up France's vineyards. Already Japanese investors had acquired interests in two Haut-Médoc wineries, Château Citran and Château Reysson, and the Château Lagrange, which had one of the largest holdings of vineyards in Bordeaux.

Theirs being a country of fierce nationalism, the French

didn't take kindly to foreign influences. The presence of American ketchup on the tables of France's restaurants was enough to draw their ire and call for laws outlawing the condiment. Bordeaux châteaux...bottled culinary accoutrements...and now the crown jewel of their French wine heritage was going to be compromised by outside influence?

Not quite.

Somebody somewhere in the French press, as Ushiyama had described, had caught wind of the deal between Takashimaya and Société Leroy; whoever it was who had heard "Japanese," "buying," "Romanée-Conti," went apoplectic and rumors spread. As the Domaine's spokesperson made clear, Takashimaya was not buying a piece of the DRC, rather a stake in Société Leroy, which merely distributed a portion of the DRC's wines, though it was a sizable allotment of the Domaine's wines. Around that time, Société Leroy was annually selling approximately $35 million worth of the Domaine's wines.

That is why Minister Nallet took it upon himself to make clear that if Takashimaya was attempting to buy a piece of the distribution in order to give the Japanese leverage to buy the vineyard, the French government would not tolerate such an intrusion. Nallet noted that French law gives the ministry the power to block any deal in which a non–European Community company acquires more than 20 percent stake in a French corporation.

Reassuring, but as it related to a sale of the Domaine, at least, it was a moot point. The "cathedral" was never in play. Ultimately, the talk of a Domaine sale was a big misunderstanding that caused a few weeks of headaches all the way around, but was of no enduring consequence to the DRC. However, Société Leroy and Takashimaya were about to cause extraordinarily serious problems for the Domaine, problems that would turn the entire Domaine against Lalou, even her only sister.

Jack Daniels and Win Wilson couldn't believe what they were hearing. If it was true, it was a bona fide scandal, and one that might very well drive an irrevocable wedge between them and the Domaine, or Aubert and Lalou, or both.

Reports were coming into their office that hundreds of bottles of the DRC wines just released that year in 1991—authentic DRC wines—were flooding the American market at prices considerably less than Wilson Daniels could offer. Frankly, the prices the wines were selling for were less than the price Wilson Daniels had paid for their own bottles direct from the Domaine.

Because the Domaine cellars it wines for three years before offering them to the market, in 1991 the '88 vintage was released. According to its contractually structured deal with the Domaine, Wilson Daniels paid the DRC up front for its exclusive allocation for the United States. Based on the prices Win and Jack and the Domaine had agreed to for the '88s, Wilson Daniels needed to sell its wines in America for baseline minimums in order to earn its profit. The Wilson Daniels price, say, for an '88 La Tâche was about $250. Now they were getting reports that bottles of the stuff were turning up all over the country for $125. So it went for the Domaine's other *grands crus*.

Wilson Daniels's typical clients, were choosing to buy the less expensive DRC wines, leaving old Win and Jack with a storeroom filled with Domaine wines they had, as it turns out, overpaid for and could not move, and therefore would not see the expected profit on their investment from. They were stuck with some three thousand cases, about half of their entire allocation, for a total loss of $2.7 million. That kind of red on the balance sheet at that point in the history of Wilson Daniels put them on the brink.

Win and Jack didn't need to be Sherlock and Watson to figure out what was happening. The Domaine wines that were out there were legit, and other than the Domaine and Wilson Daniels, there were only two other possible sources: Percy Fox, the exclusive distributor in the United Kingdom, and Société Leroy, with the distribution territory of everywhere else. To ensure adequate supply to service to its vast market, Société Leroy received by far the largest allocation, about 60 percent of the total release; Wilson Daniels got about 25 percent for the robust American sales; Percy Fox took the remaining 15 percent for the United Kingdom.

Each of the distributors negotiated their own market price with the Domaine based upon the unique circumstances and currency values of their territory. However, just as the world's financial markets traditionally rely on the U.S. stock markets, the axiom priceline for the DRC was the United States. The three distributors understood they were to take every precaution to ensure their allocations remained only in their market in order to prevent exactly the sort of disruption in the international market algebra that occurred in the United States in 1991—which, Win and Jack discovered, had been orchestrated in bad faith at their considerable expense. It didn't take long for Win and Jack to piece it together.

At that time, the Domaine sometimes sold DRC wines in mixed cases, which is to say, one bottle of Romanée-Conti, with the remaining eleven slots being some variety of the Domaine's other *grands crus*. The exact case breakdown was determined by production, which of course was predicated on the harvest. The Domaine sold mixed cases primarily to Wilson Daniels, at Win and Jack's request, as it made for a more practical buy for clients like restaurants and hotels. Assorted cases were also a way that Wilson Daniels could ensure the stable sales and market value of the Domaine's wines other than Romanée-Conti. In other words, if you want a bottle of the holy grail, you've got to buy the whole lot at the established prices.

In 1988, the assortment of each case was based on the following yields: The 4.46 acres of Romanée-Conti produced 6,438 bottles; the 12.4 acres of La Tâche delivered 20,137 bottles. Those are the two monopoles that the Domaine owns completely. The rest of the Domaine's *grands crus* come from appellations in which they own or lease parcels. The Domaine's nearly 13 acres of Romanée-St.-Vivant offered 19,346 bottles; the 8.6 acres of Richebourg produced 12,009 bottles; and the 8.6 acres Grands Échézeaux, 12,163 bottles. The Domaine's 4.6 acres of Échézeaux yielded quite a bit, 20,745 bottles; and the Domaine's sliver of Montrachet in the Côte de Beaune birthed 3,455 bottles of the Chardonnay.

Win and Jack discovered that Société Leroy that year also took receipt of mixed cases of the Domaine's '88s. Once those cases were prepared at the Domaine, Société Leroy sold a substantial number of them to an intermediary that paid roughly $1,800 per case, an exceptionally high price that covered just about half the total cost of each assortment. Win and Jack had dubbed the unofficial middlemen who sometimes appeared in unofficial distribution chains like this one "fax jockeys." Fax jockeys never physically took receipt of the wine, but rather handled everything via fax machine communications.

This particular Monsieur Jockey, Win and Jack learned, had arranged for the cases to be shipped to Switzerland, where all of the bottles of Romanée-Conti were removed from the boxes and sold to the markets of Great Britain, the United States, and

Japan. Based upon the scarcity and high demand and the currency exchanges in those countries, Monsieur Jockey made fantastic profits on those bottles alone. In Japan, according to what Win and Jack discovered, Takashimaya purchased many, if not all, of the bottles of Romanée-Conti. The other DRC wines were scattered into the world market, including the United States, and sold at bargain rates as the bottles had not been subjected to the normal necessary markups of the official distribution system.

Once Win and Jack were certain they had cracked and documented this "gray market," they called Aubert. They informed Aubert what they had learned and politely demanded he buy back the inventory of '88s that Wilson Daniels was now unable to sell and could not afford to sit on. After all, it was his co-*gérant* at the Domaine who had broken their distributing agreement and undercut them. Naturally, too, they made clear the Société Leroy was a problem. It wasn't just Wilson Daniels's supply chain and sales in the U.S. market that Société Leroy had upended, Lalou's company had caused damage to the long-term value and prestige of the Domaine's wines, and along the way exalted the wines from her own Domaine Leroy.

Société Leroy itself represented one of the largest domaines in Burgundy. It began with Monsieur Leroy's grandfather. François Leroy grew up a scrappy ward of the state and built a wine business. In 1868, he added to it with acquisitions he made from the Boillot family in Auxey-Duresses. François's son, Joseph, continued to develop the business, and then, in 1919, Henri, Lalou's father, took over. Under his direction, Société Leroy became an empire.

A key to Henri's success was that he took advantage of a loophole in trade laws between France and Germany. At the time, if French spirits exported to Germany contained an alcohol content above a certain percentage, they were subjected to steep taxes, a move intended to keep French spirits merchants at a competitive disadvantage in Germany. Leroy devised a supply chain in which he would make or buy fortified wines in France with alcohol content just below the taxable percentage, then ship it to Germany, where it was redistilled and sold with the higher alcohol content as brandy, which fetched higher prices and was free of the tariffs. The greater the volume, the larger the profit, and Monsieur Leroy's business moved in volumes of railroad cars.

With his vast profits Leroy expanded his *négociant* cellars in Auxey-Duresses, buying up massive amounts of the best wines and best vintages. It was said that Leroy's caves contained millions of bottles of the most valuable wines, and amounted to a wine library akin to the National Library. He also bought up vines in the best appellations of Burgundy: like Auxey, Meursault, Pommard, Vougeot, Chambolle, Gevry, and Vosne, smartly realizing that one day these capital investments, and the one he had made in the Domaine, would pay huge dividends.

Monsieur Leroy, who died in 1980, installed his eldest daughter, Lalou, as the head of his company in 1972, the year before she became co-*gérant* of the Domaine. Lalou was born in Paris but grew up in the vineyards. She claimed that by the age of three she already liked to be with her father as he tasted and discussed wines with clients. She loved tasting, and the smell of, wine. She studied at the Sorbonne, and at the University of Bonn in Germany. In addition to her native language, she spoke fluent German and Spanish. The only thing she loved as much as wine was mountaineering.

More than anything, she adored her father. She spent much of her life striving to impress and please him. Once, she broke off an engagement because her papa didn't care for the man. Lalou may not have been the son her father had hoped for, but she was every bit as clever as four foxes. In the male-dominated wine trade, she was determined to demonstrate to her father that she was every bit as smart and tough as any man. In all pursuits, Lalou's goal was to prove herself to be the very best, to make her father proud—whether she was climbing the Alps or scaling the Olympus of Burgundy.

Monsieur de Villaine didn't need Wilson Daniels to point out for him that Société Leroy's gray market scheme simultaneously eroded the hard-earned status of the DRC's wines and elevated those from her own domaine. The Société Leroy also had vines in many of the same parcels as the Domaine, Romanée-St.-Vivant, Grands Échézeaux, Échézeaux. With the '88s, while she depressed the value of the Domaine's wines from those appellations, the prices for Maison Leroy's bottles from those same vineyards remained constant or in fact elevated, depending on where Lalou set them.

It wasn't a matter of the wines selling. Wines from those appellations from the exalted names of Leroy and the Domaine would sell. Price was equated with prestige, and more concretely, it was the foundation for subsequent years' pricing. If the American market had paid $125 for '88 La Tâche in 1991 (when it should have been paying $250), it was unlikely the market would be willing to pay $300 in 1992 for the '89 La Tâche.

Aubert focused on the immediate business impact and the Domaine's obligation, as he saw it, to reimburse Wilson Daniels. He asked Lalou for the Société Leroy to reimburse the Domaine. She refused, maintaining she had done nothing wrong, that what happened to the wines happened to the wines. *C'est la vie.*

Aubert felt the Domaine had no choice but to file a lawsuit.

Meanwhile, he also felt that Lalou had demonstrated clearly

that she put herself and her company's interests over that of the Domaine and all of its family shareholders, including her own sister, and nieces and nephews. The whole affair made him sick with anxiety. His wife, Pamela, and his sisters would remember that they had never seen him so distraught.

Aubert reached out to Lalou's only sibling, her younger sister, Pauline, who agreed with Aubert. Or so she said. Aubert needed Pauline's support, her vote, at the next family shareholder meeting to oust Lalou and appoint her replacement.

In the days leading up to that family shareholder meeting in December 1991, Aubert's sister Marie-Hélène called their other sisters, Christine and Cecile. Typically, all three of them skipped the meetings and entrusted that their perspectives and best interests would be represented by their own sons and by Aubert and their other brother, Patrick. This time, however, Marie-Hélène realized that Aubert needed them to be there for him. She called her sisters and rallied them to attend.

The day of the vote, it was a full house at the Domaine. Everyone was there, including Pauline and Lalou. The awkwardness of what everyone knew would be voted on was in the air.

While Aubert had been assured that Pauline would cast the vote from her family's side that was necessary to make the change, he was not sure if she would have it in her to vote against her sister, to her face, in front of their children and nieces and nephews. When the matter was put up for the vote, all of the hands from the de Villaine side were raised in favor of impeaching Lalou; with her sister looking on, Pauline raised her hand for the electoral exclamation point. Without saying a word, Lalou turned and left the Domaine. Beyond the red gates, her husband was waiting for her in the car; she got in, they drove off, and that was that. For a few months anyhow.

The representative of the Leroy family who was voted to

fill Lalou's slot was Pauline's son, Charles. He lived in Switzerland. Only three months after he was appointed, while driving between Switzerland and France, he was killed in a car accident. The next time Pauline and Lalou saw one another was at the funeral. They didn't speak to each other. As Lalou would recall years later, "What was there to say?" Lalou had plenty to say, and she put it in a lawsuit.

Not long after her nephew was buried, the very next day as it was fixed in Aubert's mind, Lalou filed a countersuit against the Domaine for what amounted to slander and wrongful termination. The court ruled in favor of the Domaine, ordering Lalou to pay the damages with interest. Lalou was prohibited from ever holding a leadership role within the Domaine, and Société Leroy was stripped of its distribution rights with the Domaine.

The courts could do nothing to keep Lalou out of the Domaine's backyard. She had already acquired the Domaine Noëllat in Vosne, on the opposite side of the tiny town from the DRC.

Lalou was determined to make wines more critically acclaimed than the Domaine's. She believed she had been humiliated and felt betrayed by the Domaine, which she believed her father had saved not once, but twice: the first time by buying Chambon's shares and saving the Domaine from outside interests and the almost certain divisions that would have occurred in its wake; the second when her father insisted that the phylloxera-infested vines of 1945 be torn up and replanted. She felt she had been betrayed by her own sister, whose son took her place. It was accepted as an open secret throughout Vosne, through the Côte d'Or, that Lalou had not forgotten.

CHAPTER 16

Le Maître Chanteur

With more than two decades on the job, Inspector Pageault had lived the truisms of detective work that as a boy he heard his father discuss at the dinner table. The wild-goose chases of Monsieur PlayStation and the writer Bessanko had affirmed the one that goes, Every investigation comes with a dead-end lead. Two other bits of cop gospel he knew: No matter how good your instincts, there will be elements of an investigation you never see coming; and, Solving a case does not necessarily mean you have explained what happened. The Romanée-Conti investigation, Pageault was now beginning to realize, was one hell of a textbook example of all three.

Only fifteen minutes after Jean-Charles drove away from the cemetery, one of the members of the Police Nationale's stakeout team who had been scanning the area with night-vision binoculars drew the team's attention to movement in the vineyards, on the hillside to the southwest of the village of Chambolle-Musigny. Someone was moving on the hill, about one o'clock from the cemetery's front gate, out a quarter mile or so.

Inside one of the white vans parked nearby, Pageault turned

his binoculars on the hillside as directed. He could hardly believe what he saw: the silhouette of a man walking down the slope, moving through the vines toward the road at the bottom of the hill. The figure was on a line straight for the cemetery across the road.

Now where in the hell did you come from? he thought.

Pageault reminded the team to wait for his word and until then to remain vigilant for any other suspicious activity. If he had to make a bet on how this pickup would unfold, Pageault would have wagered a month's pay that it would involve a car or a motorcycle. In preparing for this night, the team had mapped out potential routes a pickup vehicle, or vehicles, might take, and then did their best to position officers to minimize the chances for a chase.

Pageault watched the man step from the vines, cross the narrow road, and enter the cemetery. A few minutes later, the man exited the cemetery carrying a trash bag, turned left, and began walking east along the dark road toward RN-74. Pageault watched the figure walk for about fifty yards. He—along with everyone else on the stakeout, linked by their silence on the radio—watched and waited, and listened for the sights and sounds of an approaching vehicle. An accelerating motor. Tires on gravel. A car pulling from a driveway or side road. But there was nothing.

Pageault watched the man walk for about another fifty yards.

Still nothing. None of the officers reported any other suspicious activity in the vicinity.

Seconds passed like seasons.

Finally, Pageault gave the word to move in for the arrest.

While some officers stayed put in position, in the event there indeed was another piece of the pickup that might get spooked and attempt to rush off, at least a half dozen vehicles zipped from all directions toward the man on the road with the trash bag. The police surrounded him and bathed him in spotlights. Hopping from their vehicles and approaching him with guns at the ready, they ordered him to drop the bag and raise his hands above his head. Blinking into the spotlights, the man did as he was told.

Pageault did the frisking. He couldn't wait to lay eyes on this guy. He was dressed in a black winter parka, black knit hat, like the one the person in the Colissimo security video wore, black boots, and jeans. A punk of a grown man, with a wide face, pug nose, eyes filled with a mix of resignation and "fuck off."

In the pat-down, Pageault found a headlamp, 523.75 euros, a receipt for a nearby hotel, a ticket for a train to Dijon, two pens, a pair of black gloves, and a couple of bank cards. No weapons.

Flipping through the bank cards, Pageault said, "So you're Jacques Soltys?"

The man confirmed that he was Jacques Soltys. He said he had been arrested before, that he knew how the game went from this point on and he wasn't interested in playing. If Inspector Pageault was wanting to know more, this Jacques Soltys said, the cop was wasting his time because Jacques wasn't going to say anything else until he saw the judge.

Thirty minutes later, at 8:15 p.m. on that February 12, Jacques, was deposited into a chair in a room inside the Police Nationale's Dijon station. Manu took a seat across from him and discovered that Jacques's defiance was a bluff. Jacques answered one question, then another, then another, for an hour. The more Jacques talked that night, the more Pageault didn't know what to make of him and his tale.

After the start with the elaborate maps, the couriered packages, the dead vines, not to mention the seemingly simple brilliance of the whole devious scheme to essentially seize as hostages the most valuable vines in the world—after all of the speculation, what Jacques now said by way of explanation was unbelievable because it was all so remarkably unremarkable.

Often, during an interrogation, for strategic reasons Pageault would fake incredulity, but in his initial interview of Jacques, over and over again he was genuinely astonished.

Only minutes into the interrogation, Pageault asked, "How could you be sure that the letters you sent via the postal service would arrive at the Domaine?"

It was the most basic of questions about what had appeared to be one of the most unsophisticated aspects of a crime. One minute police were convinced they were dealing with a masterful plot thoughtfully conceived and well orchestrated, and the next the clever bad guys were sending notes via regular mail. The disconnect represented one of the most confounding developments in the case for the police.

"Sometimes," Pageault said to Jacques, "the post office makes mistakes. The roads are bad. The delivery can be late."

The inspector was doing his best to goad Jacques into revealing some "truth" that the police might have been missing. Jacques's response, however, did not hint at any buried misdirection.

"I trusted the post office," Soltys said matter-of-factly, as if he were annoyed by what he perceived to be the detective's inferior intellect. "I trusted that if you post a letter in Dijon, it will get to its destination the next day, that's all. I didn't check on that. I didn't check on the postman, or whatever."

"How did you come up with the idea to extort the DRC and Vogüé?"

"I came up with the idea by myself," Jacques said. "It was all my thinking. I needed money and I found that was a way to get some. I picked those two domaines because of their reputation. I came last year to plan my project. I camped a little bit in the summer. I went into the vineyard to count the vines. Then I drew on a large piece of paper."

It was almost as if Jacques couldn't help boasting.

"Do you have some specialized expertise in viticulture? Did you ever work in the vines or study viticulture?"

"I went to the Lycée Viticole in Beaune for a while. I come from wine country. I know wine well. I was the owner of a vineyard in Champagne. I sold the grapes."

"Where did you stay when you were in the Côte?"

"Mostly in a cabin that I built in the woods."

"Where?"

"It is not relevant to your investigation. Maybe when I talk to the judge."

Inspector Pageault pulled back. For now, Jacques had said more than enough. Little by little, in broad strokes, Jacques had provided an overview of the whole crime.

The next morning, February 13, still skeptical of what Jacques had so willingly revealed, the inspectors drove the four hours to his home in the Louvois area of the Champagne region. There, in a tiny, squalid house, with curtains drawn closed and no sunlight, they found his wife, Martine, and the only child of Jacques and Martine, twenty-seven-year-old Cedric.

The front door opened from the street into the kitchen. The detectives noticed the sink filled with dirty dishes. Pageault drew Prignot's attention to a note that had been printed out on a computer and taped to the wall above a couple of bowls on the floor that said: "Be sure to feed the cats. ONLY wet food." In an adjacent room that, for lack of a better word, served as Jacques's study, they found a wall covered in black spray paint: the phrase "Nique Ta Mère" (Fuck Your Mother) punctuated by a giant swastika.

Pageault interviewed Martine and Prignot talked with Cedric.

Pageault sensed immediately that, although she was trying to pretend otherwise, Martine was overwhelmed, heartsick, and mortified. He got the impression that this was how she had felt for most of her life with Jacques. It was as if she were surprised to see them here, but also, she wasn't. Martine apologized for her appearance and for the state of the home. She said she wasn't expecting any visitors and had not yet had a chance to prepare herself for the day.

A plump woman in her late fifties, Martine had short fluffy hair that parted in the middle. She wore a housecoat and slippers and was without her false teeth, which made her even more self-conscious. There was the whiff of booze about her and something she wore smelled of urine.

Manu started off with small talk to get the basics. Martine was unemployed but used to work in a cardboard factory, which seemed fitting, for as she described her life, as much as it was with Jacques, it seemed just as flimsy. The last time she had seen her husband was fifteen days earlier. He had said he was going to Dijon. Since November 2008, when he'd returned home from prison, Martine said, he had left the house off and on for periods of two to three days. Yes, from time to time, Cedric went with him. No, the family didn't have a car. Sometimes her husband traveled around on a moped or a bicycle. She had never heard anything about Jacques going into Paris and knew nothing about any Colissimo mailings. Likewise, she hadn't seen any cardboard tubes or maps.

Pageault had no trouble believing Martine. The detective doubted she even knew what day it was.

Over with Cedric, Prignot was stunned. Cedric so readily admitted he was part of the vine crime that the inspector was caught off guard. She, too, had begun by trying to ascertain a bit of Cedric's biography. He worked as a gardener with the municipality of Épernay, planting perennials on the sides of the roads and that sort of thing. He lived in subsidized housing in Épernay and often came to his parents' place on the weekends, which was why he was there. Prignot informed him that his father had been arrested and why, and asked Cedric what he knew about it. Just like that Cedric came out with it.

"I was aware that my father wanted to blackmail vineyard owners. He's been telling me about this for years. It's an idea that came to him while he was in prison."

"Did you father tell you recently that he actually did it?"

"About a week ago," Cedric said, "I received at my home in Épernay a postcard that my father sent me from Burgundy in which he explained to me that he was expecting a success. What he wrote exactly was that in a few weeks we would know the results. I still have this postcard in my room in Épernay."

"What was your reaction to what your father was doing?"

"I know his past. All the things that he has done. They have made me depressed. I have tried to commit suicide several times. And I took a few things I should not have taken. I was on ecstasy, LSD, cocaine, heroin. I stopped by myself five years ago, without going into detox. Those products made me depressed. Ruined my health, and my ability to think."

"Besides the fact that your father was keeping you informed, did you have any part in the destruction of the vines or attempted extortion by your father?"

"Yes. I drilled the vines with him in one of the two domaines."

According to the notes Inspector Prignot kept of the interview, it was at this point she informed Cedric that considering he had just incriminated himself, she was going to stop the interview and formally place him under arrest.

She informed him he was entitled to a lawyer. Cedric did not want to consult with one. Considering the revelation that he suffered from depression and had attempted suicide—and considering the fact that Prignot was already sizing up that Cedric was awash in comorbid mental issues, and that, basically, this was all one sad mess—she asked Cedric if he wanted to see a doctor. He did not. Also, he said, he hoped to keep this news from his boss. Cedric did not want to lose his job planting flowers. The job had been a fresh start for him.

Over the course of the next hour, Cedric provided a detailed account of what he and his father had done, and when and where. Just as with Martine, Pageault and Prignot had no doubt Cedric was telling the truth.

Indeed, it was just the two of them, the father and son, behind the plot. They weren't working for anyone. Well, that wasn't entirely accurate. Jacques had been working exclusively for himself, which it was becoming clear he had been doing his whole life, at the expense of his wife and son. And Cedric, he was working for his father.

Cedric told Prignot that he had been looking forward to his cut of the money, and that was part of why he participated. Cedric said he was also afraid of his father and didn't feel like he could tell him no. He was concerned Jacques might sell away their family home, leave his mother and him for good; that his mother would be left homeless.

Although he was twenty-seven years old and carried himself like one of those rappers Prignot saw on TV, in a hoodie and baggy clothes with the tough-guy swagger, Cedric had a boy's face and a childlike mind. More than anything, the vibe Prignot got from him was—strange as it may sound—innocence and hurt. This was a kid, even though he wasn't a kid, who was hurting inside and had been for a long time. It was easy to imagine him on the side of a road, with a cigarette dangling from his mouth, ever so gently tucking a flower into the ground.

Prignot's read of the situation was that there was another reason why Cedric went along with Jacques—why he did it despite the fact, as Cedric had said to her, that he told his father he knew they were going to get caught. Prignot did not explore that unspoken other with Cedric. Clearly he had serious issues and she did not want to make things any worse. "Fragile." That's the word she would use to describe him. "Cedric," she would say, they all would say, was "fragile." She did not want to break him.

From the Soltys home, the detectives seized a computer, a hard drive, and various documents. They had Cedric's statement and Jacques's own words, among a mountain of other evidence, including the security video. They went to Cedric's apartment and retrieved the postcards and other correspondence he had described.

On their drive back to Dijon, the investigators had cause to be happy. The crime against the Domaine de la Romanée-Conti had been solved and turned out to be something far less nefarious than anyone had ever imagined. Prignot was sure they were prepared to answer all the evidentiary questions that would come with the trial and that the prosecutors would win and Jacques and Cedric would be convicted. Yet neither of the detectives felt much like celebrating. Of all the many possible conclusions the two detectives had gamed out, none of them had looked anything like this. Rather than feeling a sense of accomplishment, they left the Soltys' home feeling full of... what? The detectives were not sure how to describe what they were feeling.

On the long ride home Pageault and Prignot discussed how the plot to commit such a crime against the greatest vineyard and most storied domaine in the world began and ended in such a dark, depressing house in Champagne, with such a sad, fractured family; in a place filled with such emptiness. That was how they felt—full of emptiness. It was almost as if they had caught the emptiness in there like one would catch the germ of a cold. Ultimately, their conversation turned to Cedric and what they both believed was his true motivation—love, a son's love for his father.

Pageault and Prignot spent much of the drive back wondering just what it was that had happened to Jacques Soltys, why he grew into someone who was capable of poisoning so much.

Sometime before 1953, Antoine and Françoise Soltys emigrated from Poland to France and settled in the Champagne region, near the city of Reims. As with any immigrant story, the Soltyses moved elsewhere in the hopes of putting down roots and cultivating a better life.

Champagne's reputation as the "the drink of kings" was born of the fact that Reims was where the ancient French kings were crowned. The Reims Cathedral was to the French monarchy what Westminster Abbey was to England's royalty. The aura of that regal history and wealth bubbled from the grand champagne houses that line the cobblestone boulevards of Reims: Veuve

Clicquot, Charles de Cazanove, Ruinart, G. H. Mumm, Dom Pérignon, Taittinger.

Antoine and Françoise hoped to capture enough drops of Champagne's prosperity to make a decent living. They believed they could grow a bright future in the vines of white grapes. And indeed they did. Antoine and Françoise went to work in the vineyards and saved enough money to buy vines of their own. About an acre in Bouzy, and about an acre in Ambonnay. It wasn't much, but enough, and the vines they acquired were of a fine quality, producing *grand cru* champagne grapes. Antoine arranged a contract to sell his harvests to the champagne house of Georges Vesselle.

Antoine's vines and his two children were everything to him. Cecile was born in 1947, and then came Jacques on July 28, 1953. They were baptized into the Catholic faith, which their parents took seriously. Cecile proved herself to be an excellent student, and at church little Jacques served as an altar boy.

Antoine envisioned his vines and children growing together. He imagined a day when he would pass along the estate of his hard work to his two children and they would care for the vines; maybe Cecile and Jacques would work together in the vines with families of their own; maybe they would grow the Soltys vineyard holdings into something more. Maybe one day the Soltys name would be on one of those grand champagne houses on the cobblestone boulevard in Reims, and they would be the ones buying grapes from farmers just like him.

Those hopes would never flower. Virtually from the moment Jacques was able to make his own choices, he made it clear that he did not share his father's passion for the vines. In less than a year, teenage Jacques was expelled from the Lycée Viticole in Beaune. He was a discipline problem who mouthed off to teachers, according to his file at the school. A short while later, he entered the army, where he worked as a tank mechanic. After the service, he returned home and without a better option went to work in the vines with his father.

In the mid-1970s, while he was at a bar near Ambonnay, Jacques met Martine Richomme. Her parents were also champagne vignerons. Jacques and Martine were both in their mid-twenties. What Martine first noticed about Jacques was his striking blue eyes. She was beautiful then, and she knew that he was taken by her when he started inviting her on dates to nice restaurants and bringing her flowers. The second oldest of nine children, Martine wasn't accustomed to such attention. Martine hadn't yet taken the time to envision her future. She had not been raised to have dreams of her own.

Jacques became her life. They married in October 1978, and while she was pregnant with their first and only child, they purchased their home in Louvois, the very same home Inspectors Pageault and Prignot visited. For a very brief time, just after Cedric was born, Jacques and Martine's life together was uneventfully happy. On Sundays they would take Cedric to have dinner with Jacques's parents. Antoine and Françoise would make pizza and quiche and they would fuss over the baby. It was the sort of life Martine had wished for when she married Jacques. It was more than enough for her.

That was not enough for Jacques. Martine would not remember when or why exactly, but in the early 1980s, her husband grew steadily dissatisfied with their life, with his life, with her, with Cedric, with life in general, and especially with his work with his father in the vines. He had never enjoyed vigneron work and now he hated it. He felt that his father was taking advantage of him and not paying him enough. Jacques and Antoine were always arguing over his salary.

Jacques began to drink heavily. The more he drank, the more angry he became. The angrier he became, the more he drank. Antoine didn't like that Jacques drank so much and didn't want him coming to work in the vines when he did. That was an easy one for Jacques; he chose to drink. The family got by on Martine's salary from the cardboard factory. Which made Jacques even angrier.

One day in the mid-1980s, Jacques got it into his pickled brain that he was going to rob a bank. In the middle of a bright, sleepy day, Jacques, who was then in his mid-thirties, walked into a Banque Postale in Auxerre with a pistol. According to a French newspaper account, the "heist" was the stuff of black comedy, worthy of a French sound track of trumpet, brushes on a snare drum, and strumming guitar.

Jacques walked in, no mask, pointed the gun at the teller, and demanded whatever was in the teller's money drawer. Behind the glass, the clerk sized up Jacques and determined that pistol or not, Jacques wasn't much of a threat. The clerk put about two hundred dollars on the counter and said that's all that was in the drawer and surely it wasn't worth going to prison over. Stunned by the response, Jacques hesitated for a moment and said, "Well, give me twenty." The teller refused and told Jacques to just go away. Jacques mumbled something the clerk couldn't make out, turned and walked away. He was picked up within in a few blocks. The gun wasn't loaded. He was drunk. Without a previous record and thanks to the help of his parents, he got off with little more than a warning.

A few years later, in 1986, while Cedric was off at school and Martine was off at the factory, Jacques went off the deep end. In the middle of the day, he was so intoxicated, he urinated on the wall outside a Banque Postale and belligerently accosted people on the street. When the gendarmerie arrived to the scene, Jacques picked up a pipe and attacked them. He was arrested and this time sentenced to seven months in prison, but was out in five.

Because Jacques did not have the money to pay his legal fees and fines, the courts garnished Martine's salary. The family had barely been paying their bills as it was and now they fell into debt. Jacques's solution was to seek his version of revenge on the government, on the Banque Postale in particular. In April 1990, just before noon, no mask, he walked into the branch in St.-Michel-Mont-Mercure with a gun and walked out with about two thousand dollars. He hit another bank in another village. This time for three thousand. After that one, to calm his nerves he walked to a nearby restaurant and had a few drinks over lunch. In the summer of 1990, he got busted stealing from a Burgundian village of Puligny-Montrachet. In the summer of 1991, he was arrested for what would be his last in a series of robberies.

At his trial, Jacques told the judge, "My wife's salary has been garnished. I told several government agencies that I just could not go on anymore, but nobody takes me seriously. What can you do when you can't live on what you are earning?"

The jury was moved by the prosecutor, who felt badly for Jacques and argued, "The accused is not a bad man. A little lost and bewildered by his circumstances." Jacques was sentenced to six years, but was released earlier, sometime around 1995.

While Jacques was in prison, Antoine died at the age of eighty-two. Jacques learned of his father's death from Martine. She told him during one of her visits. His only reaction, from what she would remember, was that he lowered his head and was silent for a bit. In accordance with French law, despite what Antoine may have wished, he had no choice but to divide his estate equally between his children. Cecile got the vines in Ambonnay and Jacques got the vines in Bouzy. Within months of being released from prison, Jacques sold his father's beloved vines for about $140,000. He spent much of it on booze and dinners and nothing. Within a year and a half almost all of the money was gone.

Jacques came up with the idea to take a hostage for ransom. He hated wealthy landowners and fancied himself as something of a Robin Hood, though a Robin Hood for himself. Steal from the rich so he would not be poor. He figured there were plenty of rich landowners in Bordeaux, what with all of their châteaux and high-priced wines, and so he decided on Bordeaux. In the early spring of 1997, he traveled to Bordeaux and set up a tent in the woods around the vineyards just outside St.-Émilion. He went to a local library and pulled the registries listing the names of the prominent châteaux owners and decided upon Luc and Beatrice d'Arfeuille, whose Château La Serre was immediately outside of St.-Émilion, surrounded by their vineyards.

Jacques spent a few weeks casing the d'Arfeuilles' home and monitoring their comings and goings. It was just Luc and Beatrice in the house. They were in their sixties. In the morning, Luc left to go to work at his *négociant* business and Beatrice stayed home, mostly tending to her garden or knitting or going for walks. One day in May, just after 8 a.m., right after Luc left the house, Jacques approached the back door. It was, as he expected, open. He stepped inside. He was wearing a bandanna over his mouth and sunglasses. In one hand he had a pistol, in the other, a rifle.

Beatrice, who looked every bit the daughter of French aristocracy she was, was slender, graceful, and calm. She was making coffee and having breakfast alone. Jacques told her not to make a sound, that he was there to take her hostage; if she did as she was told he wouldn't harm her, but if she gave him any trouble, he wouldn't think twice about killing her.

He put a pipe and handcuffs on the table. Jacques explained he was going to demand from her husband half a million francs. Beatrice sensed a shyness and an indecision in Jacques, that he didn't have brutality in him—perhaps the same qualities that that first bank teller had sensed. Beatrice responded that her life was not worth that much. She asked Jacques if he wanted a cup of coffee and talked him down to three hundred thousand francs.

It just so happened that that particular morning, Luc did not go to the offices right away. He first went to a quarry to look for some stones for a château improvement project. The quarry had been especially wet and muddy that morning and Luc figured that before going to the office, he would change his mud-covered shoes. As he approached the back door, Beatrice shouted that she was being held hostage and that Luc should not come in. As Luc fled on foot into St.-Émilion, which was only about 150 yards down a dirt path, Beatrice gave several ideas to Jacques on how he might easily escape. Minutes later the police arrived.

Beatrice ran upstairs and Jacques followed her into the room at the top of the stairs. As they listened to the police enter the house, Jacques told Beatrice that if it came to it, he was going to kill her and then himself. In the time of their brief exchange, a police officer had made it to the top of the stairs undetected. Jacques fired and missed—at the trial it was a matter of debate whether Jacques was actually trying to hit the officer or accidentally pulled the trigger—but the second shot fired definitely came from one of the gendarmes and there was no question that it hit Jacques in the chest.

Lying on the floor, Jacques begged Beatrice to tell the police that he did not hurt her, which in fact Beatrice did tell the judge at the trial. Beatrice represented herself, as she didn't want to make a fuss of it all. It was largely because of her testimony and her unwillingness to press charges, because, as she said, she felt sorry for Jacques, that he was sentenced to only fifteen years, rather than the life term that would have been typical for such an offense.

Despite what Jacques would tell police, the idea to hold the vines of Burgundy hostage was not his alone. While he was serving his sentence in a high-security prison on the island of St.-Martin, his very first year there he got to talking to the wayward son of an aristocratic family. The man suggested to Jacques that if he could take the vines of Burgundy hostage as he had tried to do with Beatrice, there would be no gunplay involved, no police, no resistance of any kind; because in Burgundy, the best vineyards are owned by very quiet, very old-money aristocrats, who care a great deal about their privacy, especially when it comes to their vines. The last thing in the world they would want to do is draw attention to the fact that their vineyards are vulnerable, or even worse, that their precious *terroir* had been poisoned or contaminated in any way.

Jacques did about nine years of his fifteen-year sentence for the Bordeaux job. For the last eight, all he did was think about the Burgundy project. He began writing to Cedric about it in 2000. In those letters he also told his son that only fools work because the government only taxes away the workingman's earnings. He told his son that he was good for nothing, and ridiculed him for not having a driver's license. He advised his son that it was better to plan for big jobs, like the one he was going to do in Burgundy, and he wanted Cedric to be a part of it. He told his son to just sit tight and wait until he got out and they would do this together. He sent him magazine clippings with articles about Burgundy and told him to pay close attention to the vineyards around Vosne and Chambolle-Musigny.

Jacques was so excited about the prospects of the Burgundy project that when he was released from prison in November 2008, he went to the Côte d'Or even before he went home to see his wife and son. Doing various jobs in prison, over those nine years, Jacques had earned about eighteen thousand euros. One of the first things he bought, after the train ticket to Burgundy, was a bike to ride through the vines of the Côte de Nuits. In prison he'd researched the top domaines and top vineyards. Romanée-Conti was the obvious choice. Musigny was selected because it was another *grands crus* vineyard and, as it turned out, it was close to the area of the woods where in November 2008, Jacques had begun to build his cabin.

Jacques spent a couple of weeks in the Côte de Nuits in November. He began to count the vines for his maps and dig out the earth in the woods above Musigny for his cabin. Throughout January and February 2009, he made several trips to Burgundy to finish his cabin and collect the information for his maps. At the local hardware stores, he bought the herbicides and headlamps and various supplies he would need.

In May 2009, he and Cedric began drilling the vines of Musigny. They drilled at night between 1 a.m. and 4 a.m. By

Cedric's count they would drill about one hundred vines each night. They would scratch away the soil at the base of each vine, drill a hole, put a very tiny piece of black wire in the hole, and cover the vines. Jacques didn't want the vignerons to find the wires until he wanted them to. They did this procedure together from May 2009 until October 2009.

For more than sixteen years of Cedric's life, from the time he was ten until he was twenty-six—in other words, for most of the time Cedric had been alive—his father had been in prison. Drilling the vines and the preparations that went into it was the most time Cedric and his father had ever spent together.

Jacques would not let Cedric stay with him in his cabin. He instructed Cedric to set up a concealed tent elsewhere. During the day, he wanted Cedric to work the harvest, which he did for a few days, but then he quit because he couldn't get along with the other *vendangeurs*. Most times, during the day, while Jacques went off into Nuits-St.-Georges or Dijon, he would instruct Cedric to sit somewhere and make sure that no one stumbled upon their operation.

But there were times, Cedric told the detectives, when his father would walk with him or visit him at his tent and they would talk. Cedric was going through a rough spot with his girlfriend, Margot. She wanted to break up with him, but Cedric thought she might be the one. He really loved her. He also liked rap music and wanted his father to listen to some of his favorites. Jacques advised his son not to worry about Margot. He thought she was too common, and Jacques didn't like common people.

He was always telling his son about the aristocrat he'd met while he was in prison. Jacques loved that aristocrats were educated and he loved the way they spoke. They just had this way of speaking, he told his son. Aristocrats didn't listen to rap. They listened to Mozart and Bach. Jacques reminded his son that he had always listened to Mozart and Bach, and that he had read quite a bit, and that he had his high school baccalaureate, which he had acquired in prison. Jacques thought that if things had gone differently, if fate had gone another way, he himself would have made a fine aristocrat. Instead he was the son of Antoine and Françoise, born to lowly Polish immigrants.

He told Cedric that when they collected the money he was going to buy a church, one with an organ; he was going to wire the church with the best stereo system and he was going to learn to play Mozart and Bach. As far as the money, his plan was to give Cedric three hundred thousand euros, which would have been the take from the Musigny piece, and Jacques was going to keep the million euros from the Domaine. But Jacques had said he would hold Cedric's money in an account because Cedric didn't know how to handle money.

Jacques referred to Musigny as the "spare tire." The real thing, the special target, was Romanée-Conti. Jacques wouldn't even let Cedric work that vineyard with him; that one was his. He drilled that one all on his own. When Cedric left his father for the last time in Burgundy, it was just after the harvest, sometime in October 2009. By then, according to Cedric's guess, they had drilled thirteen hundred vines in Musigny and seven hundred vines in Romanée-Conti. To be exact, Jacques had drilled 728 vines.

As far as the poisoning of the vines, only four were injected: the two in Romanée-Conti and two in Musigny, all of which Jacques did sometime after his son left Burgundy that fall. The poisoning, the Colissimo packages, the mailings, the note in the vineyard, that was all Jacques alone. He was proud of himself for that. As Cedric would put it to police, his father said that now he was the Blackmailer and everyone was listening to him. Jacques referred to himself as "Le Maître Chanteur," the Master Singer.

Shortly after the arrests were made, with Jacques being held in a prison in Dijon, and Cedric under house arrest that allowed visits to his psychologist, the police informed the owners of Romanée-Conti and de Vogüé that both had been victimized. Until that point, other than the police and Jacques and Cedric, the only person who knew that there were two domaines involved was Pierre-Marie Guillaume. He was also the vine specialist for de Vogüé; that domaine's vineyard manager, Eric Bourgogne, had called him just as Monsieur de Villaine had done. And just as he was asked, Pierre-Marie kept the confidence of his clients.

When Monsieur de Villaine was made aware that de Vogüé had also been targeted, he suggested the owners and senior staffs meet and decide how to handle the media, should it come calling, and the legal prosecution. The domaines were in agreement on the media: Should the press call, they would be forthcoming, but they hoped no calls came. They feared publicity might inspire copycats, and, frankly, they did not want the limited poisoning to be scandalously represented as if their whole parcels had been contaminated. That could be devastating for their business.

On the legal matter, the domaines disagreed. As the information of the case came in, Monsieur de Villaine and Claire and Marie Ladoucette all supported the prosecutor's recommendation that Jacques be penalized to the fullest extent of the law. After all, this was the first case of its kind. The prosecutor, Éric Lallement, believed it was important to send a message that such crimes against the vines of Burgundy would be met with the harshest sentences. As de Vogüé's *chef de cave*, François Millet, put it, "Burgundy is a place that has been and must be free of such evil so that man can focus on the poetry of nature that God has given us, and we can focus on our responsibility to honor that." It was critical to discourage anyone and everyone from doing something like this ever again.

However, when it came to Cedric, unlike the Ladoucette sisters, Monsieur de Villaine believed that the son and the mother had both suffered enough throughout their lives because of Jacques. The Grand Monsieur wanted the most lenient sentence possible for Cedric. Monsieur de Villaine had listened very carefully to the police when they had described Cedric and Martine's life. He had paid close attention to Prignot's rendering of her time with Cedric during that first interview and in subsequent interactions.

For Monsieur de Villaine, there was something in the notion that the only way Cedric could connect with his father was in the vineyards of Burgundy. Here was this Cedric Soltys, with no family roots of his own to speak of, an *enfant* born into a godforsaken *terroir* that his own father had ignored, abandoned; willing to commit a crime, if that's what it took to be with his father, and having those moments, as dark and misguided as they were, in the vines and *terroir* of Burgundy. As perverse as it might be, the Grand Monsieur saw something divine in that, and he did not think it was necessary or just to take action against the son.

Then something else unexpected happened.

In the summer of 2010, while he was being held in the Dijon prison awaiting trial, Jacques Soltys fashioned some prison clothing into a rope and hanged himself. In the wake of his father's death, Cedric was the only one left to face trial and whatever sentence might now be deemed appropriate to send the desired message.

CHAPTER 17

A Taste

With each tick of the clock closer to 10 a.m., the New York restaurant, A Voce, filled with more of the guests who had been invited to taste the greatest, rarest, most expensive wine in the world.

All well-heeled types, they breezed in, unwound their scarves, shed their overcoats, shook hands and did the double-cheek kiss thing. Sommeliers, critics, a handful of wealthy collectors, top wholesalers, that sort. As they waited for the affair to begin, many of them looked around the room, at least in part to see who else the Domaine de la Romanée-Conti had deemed worthy to receive an invitation. The annual "unveiling" of the Domaine's newest vintage is one of most exclusive events in the world of wine. In the United States, each year, there are only two: one in San Francisco and one in New York. I had been invited to attend the New York unveiling on March 7, 2013. This was the release of the 2010 vintage, the vintage of the crime.

A Voce is on the third floor of a high-rise on Columbus Circle, at the southern edge of Central Park. The restaurant is a slick place of black marble floors, gold-colored walls and curtains, with a wide-open floor plan. The vibe of a 1940s jazz club. Huge windows provide a glorious view of the park, which on that spring day wasn't quite so glorious. The sky was gray. The weather was cold.

A Voce's main dining area is a sunken floor, about three steps below the perimeter of the room. Stationed at each of the three stairwells to the dining area was a black-tied usher to keep guests from prematurely descending the stairs and disrupting the preparations that were silently under way. Over at the far end of the sunken dining room, just in front of the massive windows, were a few long tables covered in white tablecloths. On the tables were bottles of the 2010s. The dining room was filled with maybe twenty round, white-clothed tables, each with eight place settings of eight wineglasses, one for each of the Domaine's *grands crus*. They were arranged in an arc. A handful of staff members moved from table to table oh so carefully pouring the Domaine's wines. I overheard someone whisper to one of them, "Nobody spills here."

The Grand Monsieur, dressed in a tweed blazer and thin wool tie, moved about the room for a bit. At once, he looked entirely at ease and positively uncomfortable. He mingled about the room saying hello to old friends and greeting guests. He knew better than anyone that just as everyone there wanted a taste of the Domaine's wines, everyone wanted to say hello to Monsieur Aubert de Villaine, if for no other reason than to be able to say they shook his hand. This sort of thing came with the job.

Promptly at 10 a.m., Monsieur de Villaine approached one of the stairwells and tried to enter the dining room, when he was stopped by one of the guardian-ushers.

"I'm sorry, sir. I can't let you enter until the tasting begins."

Monsieur de Villaine smiled. He gently touched the usher's shoulder and leaned into his ear. In the Grand Monsieur's quiet way, in his gravelly whisper of a voice, he said, "It's okay, I am the winemaker."

He went to the round table closest to the tables covered in the bottles of his wines, where he was joined moments later by Jack Daniels of Wilson Daniels. Jack, who can have a booming voice when he wants one, politely summoned everyone to their seats. People trickled into the dining area as reverentially as they were entering a prayer service and took seats wherever they were available.

Jack thanked everyone for coming and said how honored he and Wilson Daniels were to once again be working with the Domaine on this, their thirty-fourth release together. He then introduced Monsieur de Villaine, calling him an "aristocrat." The Grand Monsieur stood. With his hands clasped just so, he turned toward Jack, cocked an eyebrow and said, "I am a vigneron."

Monsieur de Villaine reiterated his gratitude to everyone for attending and then talked about the story of this vintage.

He talked about the challenges the weather had presented. He talked about when the harvest began, September 22, 2010, and ended, October 5. He said every vintage presents its challenges.

This vintage, he said, "we struggled with anguish and despair and moments of hope."

He made no mention of the crime.

He invited everyone to taste the wines, encouraging them to take their time. To let the wines breathe a bit. He said that after everyone was finished, in a little while, they would talk about the wines.

Everyone sat and stared at him, clearly waiting to take their cue from him.

Monsieur de Villaine said, "Well, *bon*. I've said enough. Let's taste and let the wines speak for themselves."

Although only two of the vines of Romanée-Conti were fatally poisoned, the fact that 728 were drilled with holes was quite serious, and for the Domaine, angst-inducing. Vines, especially Pinot vines, are a vulnerable plant. Drilling a vine, as with most any plant that would be pierced through by a metal drill bit, can traumatize the vine and quite possibly kill it. As Henri Roch, the Domaine's co-*gérant*, said when he was interviewed by police, he was very concerned by the attack because just the drilling of the vines was "enough to deteriorate them."

What sort of impact that trauma would have on the fruit, and in turn on the vines, Monsieur de Villaine and his staff had no way of knowing until the fruit was picked and the wine was ready; after it was aged and was bottled over three years. And finally, when it was poured in the glass and drunk.

Naturally, the wines would not have made it this far if the Grand Monsieur had sensed any reason to withhold them, but given the public scrutiny, with all of the critics, he could never know how they would react, or what they might taste that he did not. The way he put it that day, unveiling tastings like this one are "always extremely interesting and full of information for the vigneron." The unveiling is the true test.

Every one of the glasses now sitting before us was filled not just with wine, but with all of those centuries of history—the monks, the prince, his grandfather and father, the Grand Monsieur's own life and legacy. But what of the Soltyses?

"Ghost in a glass." That's how Monsieur de Villaine's nephew, Pierre de Benoist, talks about wine. Pierre was the one the Grand

Monsieur had wooed to run his own domaine in Bouzeron. He liked to say that wine was not grapes, but rather "the ghost of the grapes." Pierre was very much like his uncle that way, and in just about every other way. Like the young Aubert, Pierre, who was the youngest of Hélène-Marie's two sons, grew up in the vines and had wanted nothing to do with them. His parents' estate was in Sancerre.

Part of Pierre's resistance to becoming a vigneron was that he had grown up knowing how hard he and his family worked in their parcels of Château du Nozay, but because their vines were not in the highest classification of appellation, the Benoists' wines were considered lesser. Meanwhile, Pierre felt vignerons nearby cut corners on their wines, put quantity and profit over quality, and yet their wines received more prestige, purely because of where the vines grew. Pierre thought it was all terribly unfair. So he had left Sancerre and gone off to law school in Paris. That is where he was when his uncle called him in 1998 and asked if he would come and run Domaine A&P in Bouzeron.

After telling his uncle no several times, Pierre received a call from his great-uncle, Aubert's father, Henri, who said to him, "Please, before you make this final decision, at least come and taste the wines." So in the spring of 1999, Pierre took the train from Paris. He toured the vines and the winery, and he tasted the most recent vintage still in barrel. The minute he drank the wines he reconsidered everything he had said, and his future.

In that moment, in that taste, he tasted home. Not home in the sense of place, but in terms of where he felt he belonged existentially. Regardless of what he had been telling himself, no matter how much fun he was having in Paris, that crisp Aligoté wine reminded him that the bottom of his shoes, and the bottom of his heart, would always be covered in the dirt of the vineyards;

that he was a vigneron and the vines was where he belonged. In 2000, he accepted his uncle's offer and moved from Paris to Burgundy and went to work for A&P.

One afternoon when I visited with Pierre in Bouzeron, I found him out in the vines, stroking the canopies and talking to the plants in a loving voice. I couldn't bring myself to disturb this Vine Whisperer. When he noticed me standing a few feet away, he smiled a smile like he knew he'd been caught in a secret.

"You talk to the vines?" I said.

"Always," he answered. "They like it."

He led me to his vintage Citroën Charleston. It was filthy inside and out. Pierre knocked some plastic water bottles and a pair of muddied boots off the tattered passenger seat and told me to get in. He took me to the winery, next to his uncle Aubert and aunt Pamela's home. As he opened some of his 2007 Domaine A&P Aligoté, Pierre explained to me that it was priced well under thirty dollars. Any wine with Aubert de Villaine's name attached, from his own vines, which he oversaw and approved, could easily sell for at least double or triple that amount, but the Grand Monsieur believed it was important to price his wines such that anyone who wanted to taste a fine Burgundy, made by a vigneron, could do so.

As we drank the wines that he and his uncle Aubert had vinified, Pierre told me, "People say that wine is grapes in a glass, but I have a different view. The grapes are gone. They are no more. What's left are the juices, the souls of the grapes, the ghosts of the grapes. These souls, these ghosts, these are what we drink; their spirit infuses our own."

It was this sort of poetry and tenderness and thoughtfulness from Pierre that had people inside and outside the family wondering if he might make the best successor to Monsieur de Villaine at the Domaine.

The anointed heir apparent, Bertrand de Villaine, was quite a different personality. About the same age as Pierre, Bertrand was not raised in the vines. He first got the idea of going to work for the Domaine while he was on a job interview in Beaune. He was interviewing for a position with a security company in the late 1990s. He mentioned that he was part of the family that owned the Domaine de la Romanée-Conti. The person looked at him and said, "Well, why are you here? When you have the greatest domaine in the world, you should be there."

Within weeks, he wrote to Aubert and asked him if he could come to work at the Domaine. This was in 2000, right when Pierre was starting in Bouzeron. To demonstrate his serious commitment, Bertrand enrolled in viticulture classes and began extensive course work in oenology. Monsieur de Villaine took on Bertrand at the DRC in 2008, paying him the lowest salary of anyone on the staff and starting him at the very bottom, sweeping the floors and picking the grapes. Meanwhile, Bertrand took viticulture classes at the university in Dijon.

Bertrand is married with five children; in order to get by on his Domaine salary, his wife sold off one of her small handful of eyeglass shops.

As the family began to look for a successor to the Grand Monsieur, everyone at the Domaine knew that according to the corporation's bylaws and out of fairness, it was time for the other branch of the de Villaines to have a representative selected as co-*gérant*. Thus far, after Edmond, it had been his son, Henri, and then his son, Aubert, who served as co-*gérant* for the de Villaines. But Aubert's father, Henri, had a brother, Jean, and Jean had two sons, and they had several sons of their own, and now it was time for someone from Jean's side.

Bertrand was the only one to step forward, and so he had preference. In fact, at a December family shareholder meeting, his candidacy was approved. Yet within the family of the Domaine, there has been some concern about whether Bertrand is the right choice. Pierre is almost Aubert incarnate. He prefers to be in the vines and with his crew. He is a vigneron in the most spiritual sense of the word, believing in the soul and ghosts of grapes.

Bertrand is more of the backslapping manager who is looking toward a future of alternative distribution systems and maximizing profits. One of the first things Bertrand asked me when we met, picking grapes in La Tâche, was if I had seen *The Hangover*, that buddy comedy movie about some guys on a bachelor-party adventure in Las Vegas. After we finished picking that morning, Bertrand pointed to the corner office, overlooking the vines that belonged to Aubert, and said, "That's where I will be."

There is no doubt that Bertrand views the Domaine's wines as a sacred family legacy; there is also no doubt that he also views them as an underpriced luxury good. The question is, which does he believe more strongly? One afternoon, Bertrand told me, "It must be hard for Aubert." With great empathy and respect, he said "When he looks at me, he sees an end for him."

Perhaps, but the Grand Monsieur is concerned about much more than his own future. "If you make a bad decision on a harvest," he told me as we had a lunch and drank a 2007 Romanée-St.-Vivant, "you have one bad vintage. If you make the incorrect decision on a director, you maybe have decades of bad vintages."

Monsieur de Villaine is wise enough to realize, and honest

enough to admit, that when he first took over the Domaine there were people who thought he might not have been the most obvious choice for the job.

And so the Grand Monsieur has been in no rush to leave the Domaine. When asked about succession, he tends to change the subject. He chooses to focus on the most recent vintage, the ghosts that are before him.

Until I made my first trip to Burgundy in the fall of 2010, on assignment for *Vanity Fair* magazine to investigate the crime against the Domaine and Vogüé, I had never before drank a burgundy, at least not that I ever bothered to notice. I wasn't a wine guy. It just wasn't part of my *terroir*.

I grew up in Philadelphia. My neighborhood was cops and firefighters, nurses and hairdressers, and roofers. Lots of roofers. People sat on the steps of row homes and on stools in corner bars and they drank beer. Schmidt's was big. Brewed in town and cheap. The only people I saw drink wine with any regularity were my mother—once in a while she'd sit in the kitchen and drink a bottle of something chilled and white that my aunt Elaine brought over—and my grandfather.

Barney McGrath was a little, fiery guy. Always wore a black knit cap. Spitting image of the actor Burgess Meredith, who played the cantankerous boxing trainer with the black knit ski cap in the *Rocky* movies. Barney sat in a brown easy chair by the front window in my grandparents' row house, white radio earpiece channeling the ball game, chain-smoking Lucky Strikes and drinking red wine. Brand was Kingsport. Screw-top. Poured it in a coffee cup.

Smelled just like the Blood of Christ in the chalice I handled as an altar boy. Rubbing-alcohol strong—"hot." After a whiff of the stuff I had to shake it off. Barney "hid" his bottles from my grandmother in the turntable compartment of their 1970s multimedia center, which was the size of a small aircraft carrier. "Do your pop a favor," he'd say, "and get me my wine."

He'd polish off a bottle in no time, then stagger into the dining room—radio earpiece dangling. He'd crawl under the dining room table and go to sleep, either because he was too smashed to make it up the stairs or because he rightly figured that when—if—he got there, my grandmother would refuse to let the drunk son of a bitch into the bed.

Those were my first impressions of wine: Ladies drink chilled, soft white while they gossip in the kitchen; old men drink strong, room-temperature red to get shellacked. And those impressions didn't change much as I grew up and became a journalist. As far as I was concerned, journalists wrote about bad people who did bad things; then those journalists drank beer or whiskey to wash down all the truths they'd uncovered.

For twenty years, I worked as a magazine writer, reporting on all kinds of crime—Hollywood crime, law and order crime, military crime, terrorist crime, governmental crime—and I drank beer and whiskey. If someone ordered a bottle of wine, I'd happily drink a glass, but I didn't ask about vintage or provenance. Was it expensive? I found that interesting and, more often than not, ridiculous. Fifty bucks for a bottle of this? Wow. Hundred dollars? You have got to be kidding me.

I thought that wine critics, the idea that people were gifted with palates so superior to the rest of us mere mortals that they could taste and therefore know wine better than the rest of us; that they could know which wines were truly good and the rest of us had mouths too stupid to do the same—I thought that was absurd. Most of the time, when I found myself reading a bit of a wine review, it struck me as being as pretentious to the point of being worthless.

In preparing for my first trip to Vosne, I came upon this tasting note for 1987 Romanée-Conti by Allen Meadows, the self-anointed Burghound, who specializes in reviewing burgundies:

> *Still relatively moderate bricking. While the lovely nose is not dominated by spicy secondary fruit, dried rose petal and warm earth aromas, there is no* sous bois *in evidence, though I suspect that it will not be long before it arrives in force. Moreover, there is absolutely no trace of rot or hail that taint, as in the case with no small number of 87s. However the absence of really complete ripeness is also in evidence as the once rich flavors are beginning to lean out though neither are they tough or unpleasant, all wrapped in a finish of moderate length.*

For the connoisseurs, this sort of review might be useful. It didn't seem especially helpful for me. Was the wine any good?

There are no shortage of studies that have made a pretty persuasive case that wine criticism could be interpreted as bunk. There was the one done in 2010 by Brian Dimarco. Dimarco specializes in helping customers, collectors, and retailers choose their wines. He runs an import and wholesale company.

One day, he gathered together his staff, which included a master sommelier and some of the most knowledgeable oenophiles in the business. He put two bottles of the same $20 wine each in a brown paper bag. He told his staff they were different wines. He identified them only by price: a $10 bottle and the other a $50 bottle. He wrote the prices on the bags. He asked his staff to choose which one they liked the best. They all picked the $50 bottle. Then

Dimarco told his staff he wanted to do a second round with two more wines. He served the same wines again, only he switched the bags. Again, his staff went with the more expensive wine.

When he was a student at Harvard, Steven Levitt, coauthor of *Freakonomics*, belonged to an elite academic group, the Society of Fellows. Often they would meet and have wine tastings. Levitt had never been a wine drinker and thought much of the talk of the nuances of wine, and especially the high prices of wines, was absurd. For one of his society's wine-tasting nights, Levitt put himself in charge of the wines. He arranged a blind tasting. At his local fine wine shop he had the clerk select two bottles of what were evidently very good hundred-dollar bottles, and he also bought the cheapest bottle of the same varietal.

He filled four decanters: 1 and 2 were the expensive wines; 3 was the cheap wine; and 4 was more of one of the expensive wines. His peers gave all four wines nearly identical ratings. They did not prefer the most expensive wines. Most interesting to Levitt was the fact that his highly intelligent peers, who insisted they could tell the difference in the taste of wines, gave the same wine that was in the two different decanters the most diverse ratings.

Robin Goldstein wrote a book called *The Wine Trials.* For decades he has studied the neuroscience of wine tasting and how price impacts it. He gathered data from seventeen blind tastings, based on six hundred studies involving more than five hundred people, ranging from amateurs to expert-master sommeliers. The overwhelming data from his studies showed that overall, people liked expensive wines less than cheap wines. In his opinion, the reason people rely so much on wine experts is that we have been convinced there is some expertise beyond our own sense of taste that is required to tell us what we think tastes good and what we think is a great wine.

Perhaps the most famous example of a wine expert famously going a long way to prove the point that wine tasting is entirely subjective is from the 1976 Jugement de Paris. One of the French judges, Raymond Oliver, who was the owner of the Le Grand Vefour restaurant and had a food television show in France, had a white wine in front of him. He looked at the wine, held it up to a light to examine the color, then took a sip. He held it up again and said, "Ah, back to France." He'd just tasted a 1972 Freemark Abbey Chardonnay, which most definitely was from California.

During my first two conversations with Monsieur de Villaine, I shared with him that I'd never had a burgundy. I told him about growing up in Philadelphia and my absolute ignorance of wine and confessed my skepticism. After our second conversation, he asked me if I wanted to taste his wines.

"Sure. Thank you."

He and Jean-Charles escorted me into the cellar. We wound our way through a labyrinth of dusty bottles, lying on their sides like so many ancient, sleeping gods. After several twists and turns we entered a room where there was a large oversize wooden cask; on it there was a single candle. Jean-Charles lit the candle while Monsieur Aubert de Villaine retrieved a dusty, unlabeled bottle. He poured some into a glass and handed it to me.

I didn't know the etiquette of a wine tasting. But by then I had researched enough of the Domaine to know that whatever he was serving me was considered one of the top twenty wines in the world and certainly was among the most expensive. As I held the glass in my hand, I wondered what I would say if I tasted the wine and didn't much care for it.

Secretly wanting to hate it, or at most think it was, eh, okay, I took a drink.

I held the wine in my mouth.

Monsieur de Villaine was standing next to me. I could feel his eyes on me. Jean-Charles was standing directly on the other side of the cask from me. He didn't attempt to hide the fact that he was studying my face. I think Jean-Charles knew that I was smiling before I knew that I was smiling.

My immediate reaction, and I mean this, was I wanted to turn and hug Monsieur Aubert de Villaine. I mean, I wanted to right then and there turn and hug this old guy, squeeze him, and tell him that this wine was absolutely the most wonderful taste I had ever had in my mouth. But since I had just met Monsieur de Villaine and I didn't want him to think I was insane, I did not. I turned to him—I was now smiling to the point of laughing—and said, "This is very good." Without thinking, I found myself tilting my glass toward him.

"May I have a bit more?"

He smiled and nodded and poured me another glass.

It was a sensation more than it was a taste, I told Monsieur de Villaine. I said, "This may sound crazy to you, but when I was a kid there was a candy called Pop Rocks. It was like candied sand and when you put it in your mouth it sort of bounced around and filled your mouth. This wine is like that but it is like from heaven. It is like divine, liquefied Pop Rocks that make me feel lightheaded—the kind of happiness that I felt after I first kissed my wife."

Monsieur de Villaine smiled and said, "That is good."

I asked him what he had served me.

It was a 2008 La Tâche.

My first burgundy.

I could not wait to taste the 2010 Romanée-Conti. I sampled all of the other wines first, and saved it for last. As I drank, I could not help but think of Cedric.

After Jacques hanged himself, Monsieur de Villaine successfully persuaded the Ladoucette sisters to join with him and be represented by the same attorney. They agreed to ask the court to sentence Cedric only to probation and charge him a fine of two euros, one for each of the vineyards that had been attacked. In the French court system, civil and criminal proceedings are combined. The court agreed to the sentence Monsieur de Villaine had requested through his attorney. However, on the day of his hearing, Cedric did not show up, and the court felt obliged to sentence him to eight months in prison. The gendarmerie sent a car to Champagne to pick him up and transport him to prison. By the time of the 2010 tasting, he had been released, back on the job. I imagined him in his municipal uniform, wearing the fluorescent-yellow-colored vest, gently tucking flowers into the grass of a median while cars rushed by him.

Finally, I picked up the glass with the Romanée-Conti and I took a drink.

For a few long seconds, I closed my eyes and held the wine in my mouth. It was magnificent. It was a sensation more than it was a taste. I saw the Grand Monsieur's face. I thought of all the ghosts that were in that glass.

I thought of a sunset I had witnessed one afternoon as I stood in the quiet at the stone cross of Romanée-Conti. The sky was covered over entirely by silvery gray clouds, the way it gets before a thunderstorm, but there was no chance of rain. All was deafeningly and wonderfully calm. There was a long horizontal seam in the clouds. It was just above the forest at the top of the *côte*, right above where Jacques had built his cabin. The seam was closing, as if being mechanically cranked shut. Behind the seam, the last bit of that day's sun shone, but only through the seam. The cloud cover everywhere else was too thick. The sunlight pouring through was the color of burnt orange, brilliant, mesmerizing. It was the most intense, the most divine light I have ever seen. The scene made you think that if you ran to the top of that hill and you got off a good jump, you would be able to grab the bottom of that seam, pull yourself up, and throw a leg over into heaven and on the other side of that seam, God Almighty himself would smile in your eyes and shake your hand.

It was the sort of light that left you with no choice but to have faith, to believe.

When I opened my eyes, I found that I had welled up.

I thought of my wife. I thought of my children, my *enfants*.

I wanted to hold them and tell them that life is good; that no matter what evil there may be in the world, there is Burgundy, there is wine, there is light, there is love.

I thought about the vines in Burgundy. Soon, the first buds of spring would open into the next harvest. The Grand Monsieur would wade through the vines as he had done for the previous four decades, studying the pastoral splendor around him, looking for the clues in nature's mystery, knowing that everything and nothing was unfolding around him.

The berries would ripen. Another vintage would be born. Once again, there would be a sense that anything was possible.

This much, though, was certain: Regardless of whatever challenges were in store, natural and otherwise, the Domaine will always be the Domaine.

Acknowledgments

This book began with a magazine story I wrote for *Vanity Fair*. I am grateful to *VF*'s editor in chief, Graydon Carter, and to my story editor there, Dana Brown, for accepting that pitch and for their many years of encouragement and support.

The first I'd ever heard of the crime against the Domaine de la Romanée-Conti was from longtime friends Bryan Ignozzi and Terin Amoroso Ignozzi. During a visit to their award-winning winery, BryTer Estates, in Napa, California, Bryan said to me, "I think I've got a story idea for you." Voilà!

By the time I began reporting the book, I'd spent nearly twenty years in magazine journalism, mostly editing and reporting stories about all sorts of depressing crimes, and I was beginning to lose faith in just about everything. I went to Burgundy to report on yet another crime, but what I discovered was the poetry of grace, unwavering tenderness, and humanity. This, as much as the *terroir*, is what flavors the wines of Burgundy. I am grateful to many Burgundians for their hospitality and assistance.

There would be no book without Aubert and Pamela de Villaine. At a DRC tasting in 2012, I was blissfully stunned by the Domaine's 2009 Corton, a new addition to the Domaine's wines. Aubert was at that event, but he was busy being the gracious host and I didn't want to bother him. Later, I wrote to him that the wine was so fantastic it made me want to get up and hug him. He wrote back, "You should have hugged me. The world could use more hugs." The world could use more people like Aubert de Villaine.

I am grateful to the Grand Monsieur's staff at the "farm" for tolerating me and my terrible beginner's French, and for helping me harvest my fruit. I am particularly thankful to the charming "Dr. No," Jean-Charles Cuvelier, and to Bertrand de Villaine. Detectives Laetitia Prignot and Emmanuel Pageault not only led me through the dense woods above the *côte* and showed me Jacques Soltys's lair, they spent days walking me through their investigation.

It was not a pleasant experience for Martine Soltys to share with me the stories of her family, but she did. I am thankful for the many hours she talked with me. Those conversations would not have been possible without the assistance of her neighbors and friends, the Milesi family, proprietors of Champagne Guy Mea.

Becky Wasserman-Hone and Russell Hone and their family opened their home and hearts to me, as they do for so many newcomers to Burgundy. They helped me navigate the roads, personalities, and culture of the Côte d'Or.

Louis-Michel Liger-Belair is far more than a "character." A patient and candid source, he provided critical context and excellent perspective, and his lovely wife, Constance, was a benevolent landlord. Burgundy is fortunate to have the "General" as one of its premier ambassadors.

My interviews with Lalou Bizc-Leroy, François and Erwan Faiveley, and Domaine de Vogüé's senior team of François Millet and Jean-Luc Pepin were critical. Vigneron and bon vivant Pascal Marchand informed what I wrote. And Domaine A&P's Pierre de Benoist was extraordinarily helpful and gracious. Jack Daniels spent hours fielding my questions.

During a harvest, I met Antoine Lecompte, who ended up proving to be just about the most capable and dedicated research assistant and translator I could have hoped for. Friend and translator Dawn Erickson was brilliant all the while. I am also grateful to Whitney Woodham for translating and helping me look for my wallet.

Chris Outcalt, Luc Hatlestad, Michael Hainey, Kristen O'Neill, Roxane White, Josh Hanfling, and Christopher and Gretchen Connelly read drafts and patiently listened to me go on and on about this project. I leaned often on Brett Garfield and Jefferson Panis.

Christopher McDougall, Ben Wallace, and Bill Gifford, longtime pals and stellar authors, were willing to think out loud with me and persuade me that I could finish what I'd uncorked.

Amanda Faison joined me for a critical reporting trip to Burgundy. Because of her presence, I dodged a herd of wild boars and witnessed one of the most brilliant sunsets I have ever seen. Along the way, she just might have saved me from myself.

When I began I researching this book I was also reporting a magazine piece about Colorado governor John Hickenlooper's first year on the job. For the better part of a year I traveled between France and Colorado; I spent half of my life in Burgundy and half in Denver at the state capitol, embedded with the Hickenlooper administration. While they are wildly different cultures, to say the least, the Hickenlooper administration shared something in common with the Burgundians: Like the vignerons of the Côte d'Or, Hickenlooper and his staff weathered the elements and labored to produce something that would enrich the lives of others and live beyond their time. I had never expected to find the sacrifice and dedication that I found in Burgundy, and I certainly never expected to find that sort of thing in politics. But there they were: Burgundians vinifying wine from grapes, and the Hickenlooper bunch pressing politics into public service. I was so impressed that I went to work as a media adviser to John Hickenlooper. I am honored to call him a friend and I will forever be grateful to him and to his senior team, now my colleagues, for their patience and support: Alan Salazar, Eric Brown, Kevin Patterson, and Roxane White.

AJ, Charlie, and Sally Fairbanks of Hyde de Villaine winery made me feel at home during research in California.

Master sommelier and one of Boulder, Colorado's most knowledgeable wine merchants, Brett Zimmerman, along with Max Marriott, the regisseur/vigneron for Chapter 24 in Oregon, read my manuscript and did their very best to help me avoid errors on matters of wine.

Professor John D. Woodbridge, a historian who is as relentless as any investigative journalist I have known, is an expert on the Prince de Conti's covert activities. He patiently fielded my questions and helped me animate his own masterful research. I should note that there is no record of how and when exactly the Prince de Conti traveled to his treasonous meetings on the banks of the Seine. After speaking with Woodbridge, however, and reading his exhaustive research on the subject and applying common sense, I felt it was reasonable to conclude that the prince would have traveled by carriage, at night.

Helen Grall-Johnson, a lecturer in the French Department at the University of Denver, translated mountains of French police documents and history books.

The Denver Art Museum's exhibit on French Impressionists provided inspiration just when I needed it most.

Wine luminaries John Kapon, Allen Meadows, and Jasper Morris allowed me to tag along and answered many of my elementary questions.

My agent team of Larry Weissman and Sascha Alper made

this first book of mine happen. In other words, they helped me realize a dream.

The best editors I've ever had the good fortune of working with not only have great counsel when it comes to what's on the page, but more than that, they are good human beings with a great bedside manner. Sean Desmond, my editor at Twelve, quite literally helped keep me sane; he bought me time, and when I needed it most, he understood.

My parents, Allen and Eleanor, and my brother, Michael, and my aunt Patricia McGrath, as always, were there.

My wife, Lori, and our two sons, Jack and True, are my *terroir*.

If I have made mistakes in presenting this story, I hope they do not distract from the magic of the Domaine de la Romanée-Conti and Burgundy. My aim was to bottle the truths in such a way as to enthrall and inform, to give readers a context so that should they desire to drink a Burgundy wine they might more fully appreciate all of what is in their glass. Most of all, I aspired to produce a work that captures the essence of Burgundians, and to honor them and their sacred land.

I am deeply indebted to the work of many authors, but these books were invaluable source material: Professor John D. Woodbridge's exhaustively researched *Revolt in Prerevolutionary France: The Prince de Conti's Conspiracy Against Louis XV (1755–1757)*; George Gale's *Dying on the Vine: How Phylloxera Transformed Wine*; *Romanée-Conti* by Richard Olney; *Madame de Pompadour: A Study in Temperament* by Marcelle Tinayre; *The Pearl of the Cote* by the ultimate Burghound, Allen Meadows; *Burgundy to Champagne: The Wine Trade in Early Modern France* by Thomas Brennan; *Terroir and the Winegrower* by Jacky Rigaux; Jean-Francois Bazin's *La Romanée-Conti* (in French); and Gert Grum's *Le Domaine de la Romanée-Conti*.

Index

About the Author

Award-winning journalist MAXIMILLIAN POTTER is the senior media adviser for the governor of Colorado. He has been the executive editor of *5280: Denver's Magazine*, and a staff writer with *Premiere*, *Philadelphia*, and *GQ* magazines. Potter has been a contributing editor to *Men's Health/Best Life* and *Details*, and he has contributed to *Vanity Fair*. A fellow of the Knight Digital Media Center Multimedia Reporting and Convergence Program, Potter is a native of Philadelphia, with a BA from Allegheny College and an MSJ from Northwestern University's Medill School. He lives in Denver with his wife and two sons.